SpringerBriefs in Psychology

Series Editors
Daniel David
Raymond A. DiGiuseppe
Kristene A. Doyle

Epidemiological studies show that the prevalence of mental disorders is extremely high across the globe (World Health Organization, 2011). Moreover, and what is perhaps more concerning is the fact that, despite numerous existing evidence-based treatments for various mental disorders, more than half of those in need of specialized mental health services don't access it and/or do not have access to these treatments (Alonso et al., 2004c; Kohn, Saxena, Levav, & Saraceno, 2004; Wang et al., 2005). Thus, developing and disseminating accessible evidence-based protocols for various clinical conditions are key goals in mental health. This effort would nicely complement the efforts of the American Psychological Association (see Division 12's List of evidence-based treatments), National Institute for Health and Clinical Excellence (see NICE's Guidelines) and Cochrane Reviews (see Cochrane analyses of various clinical protocols) that identified evidence-based treatments for various clinical conditions, based on rigorous literature analyses. However, once identified, one needs a detailed published clinical protocol to deliver those treatments in research, clinical practice, and/or training (see David & Montgomery, 2011). Please submit your proposal to Series Editor Daniel David: daniel.david@ubbcluj.ro.

More information about this series at http://www.springer.com/series/10143

Lenka Ďuranová • Sandra Ohly

Persistent Work-Related Technology Use, Recovery and Well-being Processes

Focus on Supplemental Work After Hours

 Springer

Lenka Ďuranová
Business Psychology Group
University of Kassel
Kassel, Germany

Sandra Ohly
Business Psychology Group
University of Kassel
Kassel, Germany

ISSN 2192-8363   ISSN 2192-8371   (electronic)
SpringerBriefs in Psychology
ISBN 978-3-319-24757-1   ISBN 978-3-319-24759-5   (eBook)
DOI 10.1007/978-3-319-24759-5

Library of Congress Control Number: 2015950345

Springer Cham Heidelberg New York Dordrecht London

Printed on acid-free paper

Springer International Publishing AG Switzerland is part of Springer Science+Business Media (www.springer.com)

Acknowledgments

This work has been co-funded by the Social Link Project within the Loewe Program of Excellence in Research, Hessen, Germany, and an internal funding of the University of Kassel. We would like to thank the researchers of the Social Link Project for fruitful discussions and input.

Abstract

The aim of this work is to provide insight into the process of employee recovery and well-being in regard to work-related ICT use during after-hours. Therefore, we discuss (1) theories that help us to understand the determinants and outcomes of this behavior, (2) our core concepts recovery and well-being, and (3) previous empirical findings on ICT use after hours for work purposes. On the basis of literature review, we propose a new conceptual overall framework of ICT use after hours for work purposes with the focus on employee recovery and well-being processes. Thereby, we posit ICT use after hours for work purposes as potential stressor, resource, or demand (see action theory by Hacker 1998, 2003; Frese and Zapf 1994), depending on many personal and environmental factors but primarily on cognitive appraisals (see transactional model of stress by Lazarus and Folkman 1984). This three-way division enables us to propose various linear and nonlinear associations to focused outcomes. We conclude with an overall discussion on further research concerning the identified research gaps.

References

Frese, M., & Zapf, D. (1994). Action as the core of work psychology: A German approach. In M. D. Dunnette, L. M. Hough, & H. C. Triandis (Eds.), *Handbook of industrial and organizational psychology* (Vol. 4, pp. 271–340). Palo Alto: Consulting Psychologists Press.

Hacker, W. (1998). *Allgemeine Arbeitspsychologie: Psychische Regulation von Arbeitstätigkeiten*. Bern: H. Huber.

Hacker, W. (2003). Action regulation theory: A practical tool for the design of modern work processes? *European Journal of Work and Organizational Psychology, 12*(2), 105–130. doi:10.1080/13594320344000075.

Lazarus, R. S., & Folkman, S. (1984). *Stress, appraisal, and coping*. New York: Springer.

Contents

About the Authors

Lenka Ďuranová is a research associate of business psychology at the University of Kassel, Germany. She received a PhD in Literary and Cultural Studies from the Gießen University, Germany, in 2009. Her dissertation focused on motivation behind suicide. Lenka's research interests include employee well-being, occupational stress, and work attitudes. Her current research examines the role of work-related use of new technology after hours for daily fluctuations in well-being. She can be contacted under lenka.duranova@uni-kassel.de.

Sandra Ohly is a professor in business psychology at the University of Kassel, Germany, since 2010. She received her PhD from the Technical University of Braunschweig, Germany, in 2005 and her habilitation from the Goethe University Frankfurt, Germany, in 2010. Her research focuses on well-being, creativity, and proactive behavior. She is also interested in affective and motivational processes, oftentimes using diary methods. In a recent research project, she examines how smartphone use after hours relates to well-being and work-home interference. Her research has been published in the Journal of Organizational Behavior, Journal of Applied Psychology, Journal of Occupational and Organizational Psychology, and Journal of Business and Psychology. She is an associate editor of the Journal of Personnel Psychology and member of the editorial board of the Journal of Organizational Behavior, Journal of Occupational and Organizational Psychology, and Journal of Business and Psychology. She can be contacted under ohly@uni-kassel.de.

Abbreviations

ICT   Information and communication technology
TASW   Technology-assisted supplemental work (see Sect. 4.1)

Chapter 1
Introduction

Knowledge work in the 21st century has been changed considerably in comparison to the last century. In particular, the rapid development of new technology provides the greatest contribution to the major changes. Many knowledge workers can work anytime and anyplace (Davis 2002). This circumstance affects working conditions in several serious ways. As one of the effects, the use of new information and communication technology (ICT), such as smartphones and laptops, for work purposes during after-hours (labelled TASW, see Sect. 4.1) tends to increase continuously (Arlinghaus and Nachreiner 2014), giving the impression of 24/7 employee availability. This calls for changes not only in the individual organization of (non-) work time but also in HR management as it implies substantial modifications in job characteristics, such as autonomy, which, in turn, suggest serious impact on work outcomes (see Deci and Ryan 2012; Hackman and Oldham 1980).

However, up to now, we have little understanding of the implications for working individuals. On the one hand, it seems reasonable to suppose that the possibility to work using ICT when, where, how, and how long an employee likes, might have positive consequences for him or her as, for example, two knowledge workers expressed:

> It saves time ... I think it has increased my productivity. The greatest advantage is .... multitasking. I think it is a very good productivity device ... I would like them [the manufacturers] to sort of fool proof them, but, all in all – magic toy. (Matusik and Mickel 2011, p. 1015)
>
> Say there was a decision made – no Blackberrys for all employees. I would have great difficulty with that for I think I would be in situations where I would be compelled to stay at the office. I would lose flexibility. It's an advantage to the organization for I am more productive and it's an advantage to me because I have that flexibility and control. (Duxbury et al. 2014, p. 580)

© The Author(s) 2016
L. Ďuranová, S. Ohly, *Persistent Work-Related Technology Use, Recovery and Well-being Processes*, SpringerBriefs in Psychology,
DOI 10.1007/978-3-319-24759-5_1

On the other hand, there are also negative reactions to the perceived constant always-online expectations, as another employees reported:

> There is too much expectation placed on people who have these things [BlackBerrys] – you can do it. In other words you don't have a private life and things to do over the weekend. There is an expectation that comes with the privilege of having these technologies that you're always available. Since you have it you are going to do it. We expect it. There is not much choice at that point to say, 'No, actually I can't do it I have better plans'. You just can't shut the door when you want to. (Duxbury et al. 2014, p. 582)
> Sometimes while driving, I feel like instead of relaxing, I should be more productive, like calling people back, calling my customers back. Sometimes it's good that people I am working with can reach me 24 hours, 7 days a week. Sometimes, however, it might be destructive; it makes me feel I should be working more than I am. (Jarvenpaa and Lang 2005, p. 11)

In accordance with such confusing considerations of ICT use for work as positive as well as negative, existing research labeled the phenomenon of constant connectivity as *double-edged sword* (see, among others, Diaz et al. 2012; MacCormick et al. 2012). Similar views concerning the anytime and anyplace use of technology have termed its effects as *autonomy-* (Mazmanian et al. 2013) or *empowerment/ enslavement paradox* (Jarvenpaa and Lang 2005).

Obviously, the variations in individual responses to the same stimulus ('always on') need to be understood. However, although the research in this field has greatly increased recently, an overall conceptual framework is still missing. Therefore, the aim of the present work is to provide such an overall framework of antecedents and consequences of work-related ICT use during non-work time. Thereby, we focus on individuals' recovery during non-work time and on well-being as two important concepts in the work and industrial psychology (see, for example, 'the happy–productive worker hypothesis' as emphasized by Wright and Cropanzano 2000). Thus, in the following chapters we will discuss:

- theories that help us to understand the determinants and outcomes of ICT use after hours for work purposes,
- our core concepts recovery and well-being,
- previous empirical findings on ICT use after hours for work purposes,
- an overall conceptual framework of ICT use after hours for work purposes with the focus on employee recovery and well-being processes that considers both existing theoretical assumptions as well as previous empirical findings.

In developing our conceptual framework, we argue that the previous consideration of ICT use as 'double-edged' (demand/resource) is not sufficient to explain all its possible consequences for our core concepts of recovery and well-being. Therefore, we posit ICT use after hours for work purposes as potential stressor, resource, and/or demand (see action theory by Hacker 1998, 2003; Frese and Zapf 1994), depending on many personal and environmental factors, but primarily on cognitive appraisals (see transactional model of stress by Lazarus and Folkman 1984). The three-way division (stressor/resource/demand) enables to propose various linear and non-linear associations to recovery and well-being processes. We conclude with an overall discussion on further research concerning the identified research gaps.

References

Arlinghaus, A., & Nachreiner, F. (2014). Health effects of supplemental work from home in the European Union. *Chronobiology International, 31*(10), 1–8. doi:10.3109/07420528.2014.957 297.

Davis, G. B. (2002). Anytime/anyplace computing and the future of knowledge work. *Communications of the ACM, 45*(12), 67–73. doi:10.1145/585597.585617.

Deci, E. L., & Ryan, R. M. (2012). Self-determination theory. In P. A. M. Van Lange, A. W. Kruglanski, & E. T. Higgins (Eds.), *Handbook of theories of social psychology* (pp. 416–437). Los Angeles: SAGE.

Diaz, I., Chiaburu, D. S., Zimmerman, R. D., & Boswell, W. R. (2012). Communication technology: Pros and cons of constant connection to work. *Journal of Vocational Behavior, 80*(2), 500–508. doi:10.1016/j.jvb.2011.08.007.

Duxbury, L., Higgins, C., Smart, R., & Stevenson, M. (2014). Mobile technology and boundary permeability. *British Journal of Management, 25*(3), 570–588. doi:10.1111/1467-8551.12027.

Frese, M., & Zapf, D. (1994). Action as the core of work psychology: A German approach. In M. D. Dunnette, L. M. Hough, & H. C. Triandis (Eds.), *Handbook of industrial and organizational psychology* (Vol. 4, pp. 271–340). Palo Alto: Consulting Psychologists Press.

Hacker, W. (1998). *Allgemeine Arbeitspsychologie: Psychische Regulation von Arbeitstätigkeiten.* Bern: H. Huber.

Hacker, W. (2003). Action regulation theory: A practical tool for the design of modern work processes? *European Journal of Work and Organizational Psychology, 12*(2), 105–130. doi:10.1080/13594320344000075.

Hackman, J. R., & Oldham, G. R. (1980). *Work redesign.* Reading: Addison-Wesley.

Jarvenpaa, S., & Lang, K. (2005). Managing the paradoxes of mobile technology. *Information Systems Management Journal, 22*(4), 7–23.

Lazarus, R. S., & Folkman, S. (1984). *Stress, appraisal, and coping.* New York: Springer.

MacCormick, J. S., Dery, K., & Kolb, D. G. (2012). Engaged or just connected? Smartphones and employee engagement. *Organizational Dynamics, 41*(3), 194–201. doi:10.1016/j.orgdyn.2012.03.007.

Matusik, S. F., & Mickel, A. E. (2011). Embracing or embattled by converged mobile devices? Users' experiences with a contemporary connectivity technology. *Human Relations, 64*(8), 1001–1030. doi:10.1177/0018726711405552.

Mazmanian, M. A., Orlikowski, W. J., & Yates, J. (2013). The autonomy paradox: The implications of mobile email devices for knowledge professionals. *Organization Science, 24*(5), 1337–1357. doi:10.1287/orsc.1120.0806.

Wright, T. A., & Cropanzano, R. (2000). Psychological well-being and job satisfaction as predictors of job performance. *Journal of Occupational Health Psychology, 5*(1), 84–94. doi:10.1037/1076-8998.5.1.84.

Chapter 2
Theoretical Background

The aim of this chapter is to provide an extensive overview of existing theories, which can be used by conceptualizing an overall framework of antecedents and consequences of work-related ICT use during after-hours (see Chap. 5). Thus, in the following we describe the border and boundary theory (see, among others, Ashforth et al. 2000; Clark 2000), the social learning theory (Bandura 1965, 1977, 1986; Bandura and Walters 1963), the self-determination theory (SDT; Deci and Ryan 2000), the conservation of resources (COR) theory (Hobfoll 1989, 2011), the transactional model of stress (Lazarus and Folkman 1984), the job demands-resources (JD-R) model (Bakker and Demerouti 2007; Demerouti et al. 2001), the challenge-hindrance framework (Cavanaugh et al. 2000; LePine et al. 2005), and the action regulation theory (Hacker 1998, 2003; Frese and Zapf 1994). Thereby, we already offer exemplary references to the ICT use during non-work hours for work purposes. If available, we list examples of previous research and, finally, we draft our initial assumptions on antecedents and consequences of supplemental ICT work.

2.1 Border and Boundary Theory

Work-family border theory (Clark 2000) focuses on explaining the work-family balance. It concerns the physical (e.g., gates), temporal (e.g., working hours), and psychological (e.g., thinking patterns) borders associated with work and family roles, whereby borders are the "lines of demarcation between domains" (Clark 2000, p. 751). Furthermore, this theory provides a framework to facilitate the balance between them.

According to the theory, working people are daily 'border-crossers' as they transit between work and home. The major proposition is that despite being two different domains, work and family are interconnected and mutually influencing each other. Furthermore, people are affected by their environments, while likewise they

© The Author(s) 2016
L. Ďuranová, S. Ohly, *Persistent Work-Related Technology Use, Recovery and Well-being Processes*, SpringerBriefs in Psychology, DOI 10.1007/978-3-319-24759-5_2

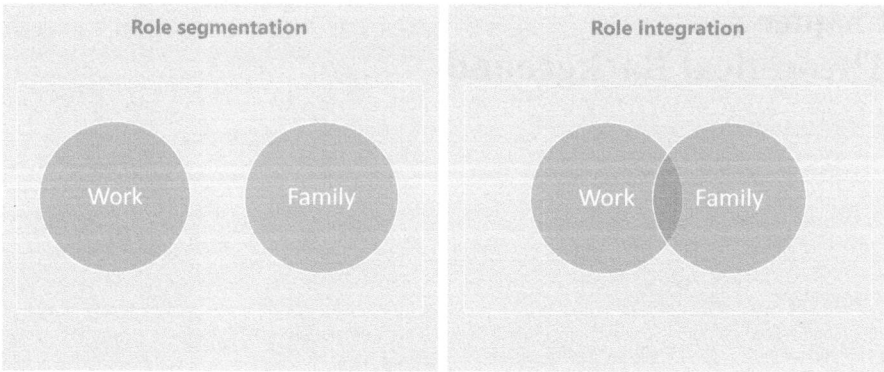

Fig. 2.1 Role segmentation versus role integration according to the border and boundary theory

also affect them. The main difference between the domains can often be noted in their cultures, whereby work and home can create quite different expectations about rules, thought patterns, and behaviors, which, in turn, shape behaviors. Accordingly, the degree of culture differences has consequences for the daily transition effort. For some people, the contrast between the work and family role is only marginal ('similar domains') and the daily transitions are not demanding, whereas those who perceive high contrast between the domains ('different domains') experience daily transitions as more demanding and, thus, their work-family balance is impaired. Other contextual determinants of border management are the social actors, such as the 'border-keepers'–at work they are mostly the supervisors and at home the spouses–who constrain the flexibility of border-crossers to deal with conflicting cross-role demands.

In fact, individuals vary in terms of managing their border-crossing ('role blurring') to attain a desired balance, which refers to "satisfaction and good functioning at work and at home, with a minimum of role conflict" (Clark 2000, p. 756). They differ in drawing the line between home and work on a continuum, ranging from high integration to high segmentation (Nippert-Eng 1996). Figure 2.1 illustrates the overlap between work and home domain in dependence on segmenting versus integrating. Thus, some individuals (labeled *integrators*) tend to integrate these domains, while others (labeled *segmentors*) prefer to segment them and prevent role blurring. The integrators may probably prefer to have an office at home and do their personal business at work, whereby segmentors prefer to separate their home and work roles both spatially and in time. In our ICT-related context, individuals who prefer segmenting are less likely to use communication technologies for work purposes after hours, whereas individuals who prefer integrating are more likely to do so. For example, the management of e-mail accounts, phone numbers, or mobile devices may represent the preference for segmenting versus integrating. Thus, high segmenting employees may not check their e-mails on their private devices or, at least, they may deactivate the push notifications during after-hours. Furthermore, they may not use the same phone number or even the same mobile device for private and work purposes (see, among others, Battard and Mangematin 2013; Collins and Cox 2014; Sayah 2013), or do not use ICT for work purposes at a specific time as the following employee:

> I think this is a question of how you structure your day-to-day work. Because I don't need to answer the phone if I don't want to. I don't have to constantly read my emails ... This constant availability is not necessary. Well, I often don't carry my smartphone with me at the weekend or I even switch it off (I3) (Sayah 2013, p. 187)

Based on the theory, one of the potential antecedents of integration versus segmentation preference is *the role identification*. It refers to higher likelihood of integrating the role with one's other roles (such as work role with family role). Accordingly, the greater the work role identification, the more likely employees may create more flexible and permeable boundaries between work and family roles to use ICT for work purposes after hours.

Still, not only the individual preference is constitutive for the actual integrating or segmenting behavior. The *characteristics of borders* can influence it too. For employees, integration is not easy if the borders are *strong*–impermeable, inflexible and not allowing blending–; such as when a spouse insists that the person does not work from home in the evenings or the supervisors sanction the use of work time for private issues. In contrast, *weak*–permeable, flexible and blending–borders impede segmenting; for example, when a spouse works in the evenings from home, or it is customary to attend to private issues at work. Clark (2000) posits improvement of work-family balance in the combination of similar domains with weak borders or different domains with strong borders.

Similarly to the border theory, the *boundary theory* (Ashforth et al. 2000) addresses daily role transitions and boundary management. It is not as broad as border theory, as it focuses mainly on the psychological movement between roles (*cognitive boundaries*). Physical boundaries (e.g., gates) are not at the focal point and temporal boundaries (e.g., working hours) are neglected. However, the boundary theory considers the transitions beyond the work-home domains additionally and, in this regard, has a broader scope than the border theory. In this respect, it includes also everyday transitions between domains involving work-home transitions, at-work transitions as well as work-another place transitions.

In the previous research on boundary management, Olson-Buchanan and Boswell (2006) examined its consequences for work-life conflict among non-academic university staff employees. In accordance with the theory, the results suggest that role identification is related to role integration. Furthermore, high integrating employees who used communication technologies during non-work time reported more work-life conflict. Moreover, Park and Jex (2011) found mediating effects of employees' boundary creation around ICT use on the relationship between individual factors (segmentation preference and work role identification) and work-family interference among office workers.

In brief, work-family border theory addresses how work-family boundary management influences work-family balance, and boundary theory focuses on how individuals create or negate daily boundaries across work and non-work domains. According to both theories, employees differ in the creation, maintenance, and modification of boundaries between the work and home domains. The individual characteristics of employees (e.g., integration preference, role identification) and contextual factors (e.g., similarity between the roles, border permeability) can serve as the reasons of role crossing. According to work-family border theory and boundary theory, it can be supposed that work-related ICT use after hours is associated with role blurring, which, in turn, may lead to higher levels of work-to-family conflict. This effect may be moderated by individuals' segmentation/integration preference. The assumptions from border and boundary theory will be integrated in the theoretical framework below.

2.2   Social Learning Theory

Social learning theory (Bandura 1965, 1977, 1986; Bandura and Walters 1963) assumes that we learn behavior within a social context by observing relevant others and imitating them. The emphasis of the theory is placed on the role of *vicarious experience* (see also Fig. 2.2). Thus, individuals acquire and imitate certain behavior when someone else appears to have been rewarded for it, and they fail to show some behavior when someone else was seemingly punished for it. This ability of people to learn by observing others enables them to avoid errors. According to Bandura (1977), whether an individual shows a certain observed behavior in a particular situation depends at first on the previous experiences of this behavior, both of the person and of others. The second criterion is the success of this behavior in the past. The thirdly crucial antecedent of modelling is the estimated likelihood of reward for this behavior. Finally yet importantly, the current situational factors—such as cognitive, social, and environmental factors—may play a role for acquiring a behavior after observing.

In the work context, some role models, such as supervisors and colleagues, are more appropriate for modelling than others. Employees learn primarily by imitating their important social referents, especially when they see that certain behaviors are rewarded. In particular, supervisors make potent models, because employees per-

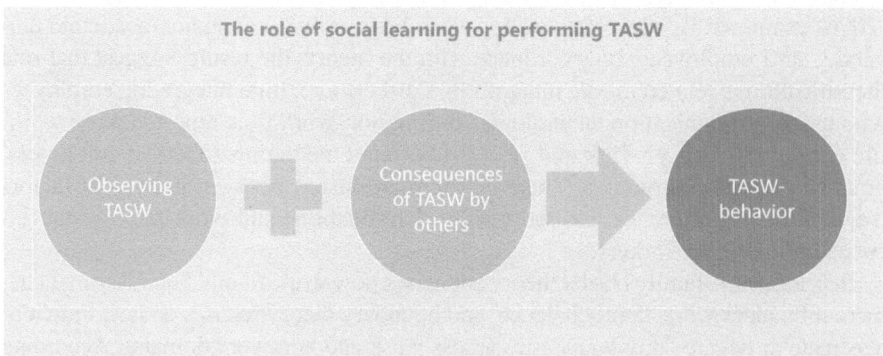

Fig. 2.2 Some antecedents of TASW according to the social learning theory (Note: TASW = technology-assisted supplemental work; see Sect. 4.1)

ceive them as respected and authoritative figures. Therefore, if supervisors use ICT for work purposes during after-hours, it will have an impact on the attitudes and behavior of their employees toward ICT use. Thus, these employees will show ICT-related supplemental work behavior in order to increase their chances of obtaining positive reinforcement from the supervisors. Meanwhile, some supervisors seem to know their impact on employees, as the following statement shows:

> He [my boss] works in New York [and] in an effort to try and role model good work-life balance, he now no longer e-mails people on weekends. And so you'll never get an email from [business head] on the weekend. And then hopefully he shows his people that I don't expect you to be working on weekends (MacCormick et al. 2012).

Furthermore, behaviors of important social referents (e.g., such colleagues) with whom employees can easily identify with are likely to be copied. In particular, the likelihood of employees' ICT use after hours will be higher after observing colleagues' ICT-related supplemental behavior (e.g., reading and sending e-mails after hours, taking laptop and work home) and it will be even more enhanced after observing colleagues' reinforcement for it (e.g., acknowledgment from the supervisor or clients).

Taken together, social learning theory posits that human behavior is learned observationally and vicariously from appropriate role models. Therefore, it can be supposed that employees' ICT use is similar to their subjective perception of ICT use of their important social referents, such as supervisors and colleagues. Park et al. (2011) have already found that the segmentation norm within a work group is negatively related to ICT use for work purposes outside of work. Furthermore, the role of social influence has also been integrated in the well-known unified theory of acceptance and use of technology (UTAUT; Venkatesh et al. 2003). Thus, we would expect that the individual norms and attitudes toward ICT use, in the course of time, become more similar to those of their organization, department, work group, or any other significant social groups. The assumptions from the social learning theory will be integrated in our conceptual framework in Chap. 5.

2.3   Self-Determination Theory

A further useful motivational theory is the *self-determination theory* (SDT) of Deci and Ryan (2000). It posits the existence of three *universal psychological needs*–needs for competence, autonomy (labeled also as 'self-determination'), and relatedness. Furthermore, the satisfaction of the needs is proposed to be associated with growth, integrity, and psychological well-being. These needs are innate and need not be learned. Therefore, all individuals are predetermined to search for psychological growth, being naturally actively engaged and socially connected. These basic needs are essential for understanding the motivation of behavior.

As can be seen in Fig. 2.3, SDT describes human behaviors along a continuum between self-determination and non-self-determination. Furthermore, it distinguishes between several types of motivation with different, underlying regulatory

Behavior	Locus of Causality	Motivation	Regulation	Regulatory Process	Item Examples
Nonself-determined	Impersonal	Amotivation	Lack of Motivation	Non-regulation	"I don't, because I really feel that I'm wasting my time at work."
	External	Extrinsic	Controlled	External	"Because I risk losing my job if I don't put enough effort in it."
	Somewhat External	Extrinsic	Moderately Controlled	Introjected	"Because otherwise I will feel bad about myself."
	Somewhat Internal	Extrinsic	Moderately Autonomous	Identified	"Because putting efforts in this job has personal significance to me."
	Internal	Extrinsic	Autonomous	Integrated	"Because it has become a fundamental part of who I am."*
Self-determined	Internal	Intrinsic	Inherently Autonomous	Intrinsic	"Because I have fun doing my job."

Fig. 2.3 The core concepts of the self-determination theory (SDT) (Source: Adapted from Deci and Ryan 2000; Gagné and Deci 2005. Note: The item examples were taken from the *Multidimensional Work Motivation Scale* by Gagné et al. 2015. Thereby, participants are asked: "Why do you or would you put efforts into your current job?" *This item stems from Tremblay et al. 2009)

processes and loci of causality along this continuum. Within motivation, the fullest self-determined behavior is intrinsically motivated, the opposite behavior is amotivated, and between these two poles, the extrinsically motivated behavior is located. People are *intrinsically motivated* when they perform an activity because they find it interesting and do not require any reinforcement. For example, intrinsically motivated employees use ICT for work purposes in their leisure time because they enjoy doing it (see the item example by Gagné et al. 2015 in Fig. 2.3). Intrinsically motivated behaviors are based on the three universal needs, but particularly on the need for competence and autonomy. Hence, intrinsic motivation is proposed to be facilitated by conditions that result in psychological need satisfaction and undermined by conditions that thwart need satisfaction. *Amotivation* means a complete lack of motivation or intention to behave. *Extrinsic motivation* refers to performing an activity because of its expected instrumental consequences. SDT proposes that extrinsic motivation varies with respect to the degree to which behavior is self-determined. It distinguishes two types of *controlled extrinsic motivation*: external and introjected regulation; and two types of *autonomous extrinsic motivation*: identified and integrated regulation. *Externally regulated behavior* is controlled by external instrumental consequences, such as tangible rewards or threats of punishments. For example, externally motivated employees use ICT for work purposes in their leisure time because they expect financial rewards after doing it or avoid being criticized by their colleagues or supervisors after not doing it. *Introjected regulated behavior* is controlled by only partially internalized instrumental consequences. In other words, introjection refers to regulations within the person, but they still do not become a part of the one's self. For example, introjected motivated employees use ICT for work purposes in their leisure time in order to obtain feelings such as pride of themselves or avoid guilt and shame by not doing it. *Identified regulated behavior* refers to the identification with the target behavior and its values. People who identify with the values of some behavior accept it as a part of their self rather than by introjection. For example, employees identified with the importance of supplemental work for their career advancement would use ICT for work purposes in their leisure time more volitionally. *Integrated regulated behavior* represents the most complete and effective internalized form of extrinsic motivation. The resulting behavior is autonomous and fully volitional, although still extrinsic. For example, integrated motivated employees use ICT for work purposes in their leisure time because this behavior fully correspondents with their personal values. Within *loci of causality*, SDT allocates the various motivations to various loci of causality along the continuum from *external* to *internal*. Amotivation has an *impersonal* causality locus.

Thus, SDT has a special motivational approach. Specifically, the self-determination continuum is located between amotivation and autonomous motivation. Accordingly, it distinguishes between several types of motivation with different, underlying regulatory processes. Based on SDT, we may suppose, at first, that the opportunity for work-related usage of ICT during non-work time leads to higher perceived work autonomy and this, in turn, facilitates workers' well-being, as this employee reported:

> It's just hugely important to me that I'm there three or four nights to see [my two year old] before he goes to bed. I leave the office at 4 p.m., go home, spend a bit of time with my son, have dinner with my wife . . . then pull out the BlackBerry and work until 10 or 11 p.m. (MacCormick et al. 2012, emphasis in original);

and another manager said:

> I can connect when and where I want ... It means I am a lot less stressed because I can go home when I need to [...] I can stay on top of what needs to be done (MacCormick et al. 2012, emphasis in original).

Secondly, we may expect that employees' different motivations for work-related ICT use after hours relate to different outcomes of ICT use. In previous research, Ohly and Latour (2014) have already examined the relationship between autonomous versus controlled motivation for smartphone usage in the evening and well-being in the evening. The results are in line with the assumption that the motivation of individuals to use their smartphone for work in the evening matters for their well-being. The details of the assumptions will be integrated in the theoretical framework below.

2.4   Conservation of Resources Theory

A further theory that may be helpful in explaining antecedents and consequences of ICT use during after-work hours is the *conservation of resources* (COR) theory (Hobfoll 1989, 2011). Hobfoll (2011) termed COR theory as a "stress and motivational theory" (p. 116). This resource-oriented theory is based on the central assumption that people strive to maintain, defend, and build resources they value, and they

Fig. 2.4 The first principle
of the COR theory:
resource loss more salient
than resource gain

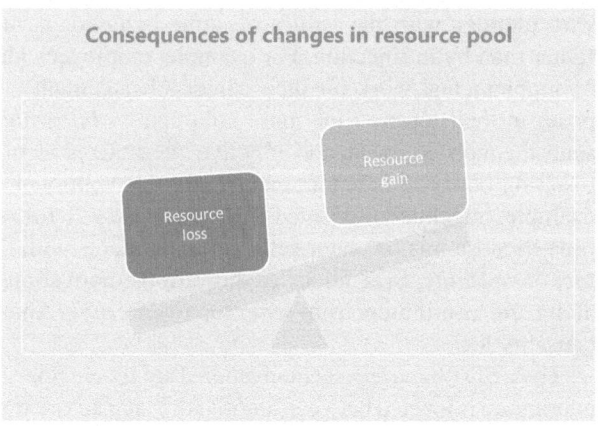

experience psychological stress when these resources are threatened, lost, or not gained after investing in them (Hobfoll 1989). Thus, people pursue steadily conservation of resources, and their actual loss, perceived loss, as well as lack of gain produce stress. Hobfoll (1989) defined resources as those objects (e.g., tools for work), personal characteristics (e.g., self-efficacy), conditions (e.g., seniority at work), or energies (e.g., knowledge, time, and vitality) "that are valued by the individual or that serve as a means for attainment of these objects, personal characteristics, conditions, or energies" (p. 516).

Hobfoll (2011) proposes two important principles in the COR theory. The first is that "*resource loss is disproportionately more salient than resource gain*" (Hobfoll 2011, p. 117, emphasis in original). Thus, the perceived resource loss is central in the stress process (see also Fig. 2.4). Second, individuals *have to invest* their resources in order to defend them, rest from their losses, and build new resources. People who have more resources are less vulnerable to their losses and more capable of building new resources. People with lack of resources are most vulnerable to additional losses, leading, in turn, even to loss spirals. According to the conservation of resource model (Hobfoll 1989), when people are confronted with stressors, they seek to minimize the net loss of specific resources, for example, by employing other resources. When people are not currently confronted with stressors, they seek to build resource spillovers in order to avoid future losses. Therefore, they invest resources to enlarge their *resource pool*. Furthermore, resources create other resources, and such resource gain cycles are critical to workplaces as well as to work-family interactions (Hobfoll 2011). On the other hand, when resource investments do not lead to the expected return, individuals will experience these as a loss, and, in turn, as stressful. Resource gain- and loss cycles "occur in chronically stressful conditions, or where individuals or organizations are resource poor and any major stressor occurs" (Hobfoll 2011, p. 118). Moreover, COR theory (Hobfoll 1989) emphasizes subjective components of loss, the *appraisal* (Lazarus and Folkman 1984). Accordingly, the possible strategies to conserve resources in context of resource loss are, for example: reinterpreting threat as a challenge, reevaluating the value of resources that are threatened or that have been lost, altering the interpretation of events and their consequences, and devaluating lost resources. The mentioned strategies may be integrated in the theoretical framework below.

In short, COR theory posits that individuals are constantly trying to maintain and gain resources and avoid their losses. Building on COR theory, we may assume that people use ICT for work purposes after hours, for example, in order to manage resources that were lost or threatened at work. Thus, they put in extra compensatory effort to avoid stress. Consequently, they invest resources, such as leisure time, energy, laptop, and knowledge in order to deal with actual loss, protect against resource loss (e.g., the perceived self-efficacy under time pressure), or gain new resources (e.g., time and control). As an executive reported concerning the mobile-enabled work:

> There is an intrusion, but ... it is controllable and compensated for with true efficiency and true tranquility. This tool allows a better distribution of one's time (MacCormick et al. 2012, emphasis in original)

However, the exposure to an extra compensatory effort too long or too frequently causes high physiological (e.g., higher vulnerability to stress and diseases) and psychological (e.g., altered emotional states and cognitive changes) costs, that may result in emotional exhaustion (Hobfoll and Freedy 1993; Gorgievski and Hobfoll 2008). Thus, people who avoid using ICT for work purposes after hours may do it, for example, in order to conserve their resources after demanding workdays. They refill their energy reserves by detaching and restoring from the daily job demands.

Overall, according to COR theory, it can be assumed that work-related ICT use after hours is associated with maintaining (e.g., self-efficacy) and gaining resources (e.g., time) as well as with their losses (e.g., psychological detachment), which, in turn, may lead to positive as well as negative consequences with regard to well-being of working individuals. The potential third variables will be discussed in the theoretical framework below. Furthermore, in according to the first principle of the COR theory, the negatively consequences may be stronger perceived as the positive consequences by the employees. In this context, Richardson and Thompson (2012) as well as Ward and Steptoe-Warren (2013) have already provided an empirical evidence of the principle that resource loss is more salient than resource gain as they showed that the mediating effect of psychological detachment on the relationship between ICT use during non-work time (e.g., evenings, weekends, and vacation) and work-family conflict was greater than the mediating effect of job control.

2.5   Transactional Model of Stress

A further cognitive theory that focuses on explaining subjective stress development is the influential *transactional model of stress* (Lazarus and Folkman 1984). The main components of this stress theory are the appraisal and coping processes.

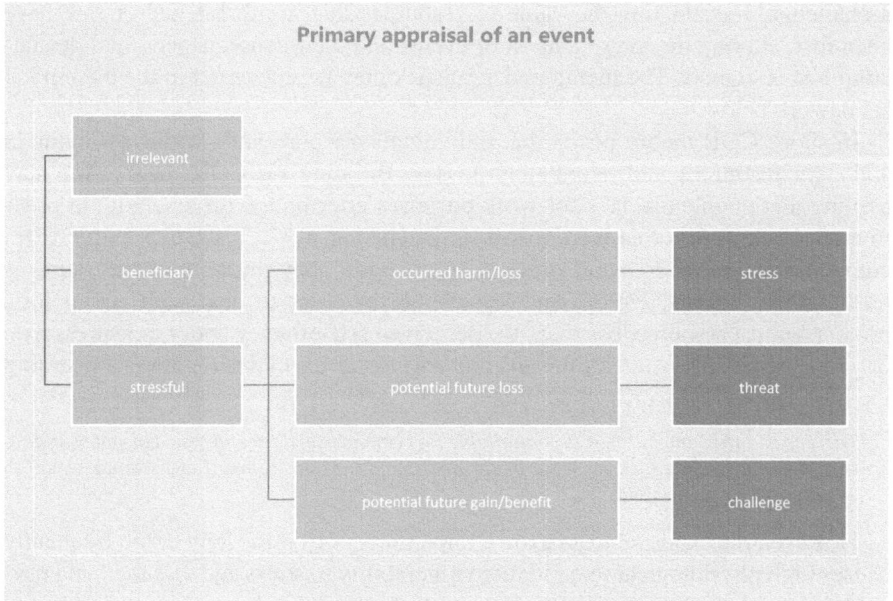

Fig. 2.5 Process of primary appraising according to the transactional model of stress

According to the transactional model of stress, stress emerges from the cognitive appraisal of a situation. Cognitive *appraising* is the "process of categorizing an encounter, and its various facets, with respect to its significance for well-being" (Lazarus and Folkman 1984, p. 31). Lazarus (2006) prefers the term *appraisal* over the often used term 'perception' because appraisal includes an automatic process of evaluating the personal significance of a situation for an individual's well-being. This automatic evaluation is not implied when using the term perception. The core assumption of the appraisal concept is that people are continuously evaluating their environment regarding its effects on their well-being. Thereby, the term 'appraisal' denotes the product of evaluation, whereas the verb 'appraise' is recommended to label the process of evaluation (Lazarus 2006, p. 57).

Cognitive appraisal takes place in two interdependent steps: During the so-called *primary appraising*, an individual judges a situation as irrelevant, beneficiary, or stressful (see Fig. 2.5); for example, in regard to an individual's well-being. Only the appraisal of an event as stressful is relevant for the development of subjective stress. In this case, three potential stressful situations are distinguished: harm/loss, threat, or challenge. *Harm/loss* occurs after stressful situations when the outcome is judged as a damage or loss. Threat and challenge appraisals occur before stressful situations when something valuable is at stake. For *threat appraisal* to occur the focus is on potential future loss; for *challenge appraisal*, the focus is on potential future gain or benefit (Lazarus and Folkman 1984). In the context of work-related ICT use outside of scheduled work hours, this behavior may be appraised as a harm or loss by perceiving impaired well-being after doing it. It may be appraised as a threat by perceiving impaired well-being while doing it. Third, appraising this supplemental work behavior through ICT as challenge may occur if the necessary overcoming of certain obstacles by doing it is connoted positively as well as negatively (e.g., when an employee expects reduction of time pressure next day even though he or she will also perceive strain by doing it).

During *secondary appraising*, an individual also judges his or her coping resources; for example, skills or social support that might help in dealing with the stressful situation. Coping, as another important concept in the model of Lazarus and Folkman (1984), is defined as "constantly changing cognitive and behavioral efforts to manage specific external and/or internal demands that are appraised as taxing or exceeding the resources of the person" (p. 141). The authors have distinguished between problem-focused and emotion-focused coping style. Thereby, *problem-focused coping* refers to reduction of stress potential of a certain situation by changing the current situation with the result that the source of stress will be handled or altered. In our context of supplemental work after hours, in order to avoid a work-home conflict, some working parents may do it only after their children have gone to sleep. *Emotion-focused coping* may be manifested, for example, by avoiding the certain situation or changing the meaning of it. Thus, an employee who is threatened by both high workload and home demands at the same time may see his or her work role currently as more important than the home role, and thereby change the meaning of the potential stressful supplemental working at home.

After the appraisal processes follows *re-appraisal*, which means a new appraisal, a re-evaluation of the potential stressful event. This re-evaluation may result in re-appraising a threat to a challenge or even to being irrelevant. Thus, the supplemental work after hours through ICT may be first appraised as a threat, but after evaluating sufficient coping mechanism it can be re-appraised as a challenge.

In previous research on technostress, the transactional model of stress was already applied in several studies (see, among others, Fischer and Riedl 2015; Lei and Ngai 2014; Yin et al. 2014). For example, Lei and Ngai (2014) propose that technostress challenge appraisal would generally lead to positive outcomes (like work performance), whereas the negative appraisal on technostress would generally lead to negative outcomes (like strain).

Taken together, the transactional model of stress states that stress emerges from a transaction between a person and the situation and, therefore, not every potential stressful situation will be appraised in the same way by every individual. These assumptions will be integrated in the theoretical framework below.

2.6   Job Demands-Resources Model

The *job demands-resources* (JD-R) model (Bakker and Demerouti 2007; Demerouti et al. 2001) focusses on the development of stress and work engagement. It was developed to predict employee burnout and engagement. Demerouti et al. (2001)

consider stress in the tradition of the transactional stress model by Lazarus and Folkman (1984) as a "disruption of the equilibrium of the cognitive-emotional-environmental system by external factors" (p. 501). External factors may lead to stress only under certain circumstances (like by low performance capacities). Under different circumstances (for example, when coping skills are high), they may also lead to well-being. Therefore, the authors prefer to use the term 'job demand' over the term 'stressor' for these external factors. They use the term *job stressor* "only when an external factor has the potential to exert a negative influence on most people in most situations" (Demerouti et al. 2001, p. 501). Thus, based on the JD-R model, the work-related ICT use during after-hours would be termed stressor only when it would mostly lead to decreasing in recovery and well-being for most employees. In subsequent work in JD-R research, the focus was on job demands, and the stressors were neglected.

In general, the JD-R model categorizes working conditions into two broad categories: job demands and job resources. *Job demands* are defined as physical, social, and organizational job characteristics that require physical or psychical (cognitive and emotional) effort or skills and bring about costs (e.g., exhaustion). The greater the sympathetic activation or subjective effort, the greater the physiological costs for the employee. Examples of demands are high work pressure, time pressure, workload, an unfavorable physical environment (e.g., noise, heat; see, among others, Demerouti et al. 2001; Bakker et al. 2003), and demanding interactions with clients. In the context of ICT use for work purposes during non-work time, this behavior may impair the recovery process of employees as it increases the overall amount of their daily workload, as noted by a principal at a small private equity group:

> So, at what point of your day does the workday end? This tool makes it difficult for that workday to end. I mean, I think, there's no doubt about it that my day doesn't really come to an end until I go to bed, right? . . . I think there's kind of a long-term negative impact because I don't think we ever get away enough if we're constantly using this [the BlackBerry]. (Mazmanian et al. 2013, p. 1346)

Further, Bakker and Demerouti (2007) assume job demands not to have negative effects under all circumstances. However, according to Meijman and Mulder (1998), they propose that the demands "may turn into job stressors when meeting those demands requires high effort from which the employee has not adequately recovered" (Bakker and Demerouti 2007, p. 312). Therefore, the ICT use for work purposes during non-work time may be termed a *stressor* when the sympathetic activation through the supplemental work hinders the subjectively needed level of recovery from work, such as a senior associate in a law firm experienced:

> I was just on it all the time. Not because I had more work to do. I was bringing my work home with me more, even if work wasn't actual legal work. I was just on this thing more, so my brain was working more, I wouldn't sleep as much and was have trouble sleeping. (Mazmanian et al. 2013, p. 1347)

Moreover, JD-R assumes an important role of resources in the detrimental demand-strain process. *Job resources* are described as health-protecting aspects of the job, which may help in achieving work goals, decreasing job demands, and enabling personal growth and development. Thus, ICT use for work purposes after hours may serve as a resource, for example, by decreasing job demands such as time pressure.

Empirical studies found support for the main model assumptions (see Bakker and Demerouti 2007; Demerouti et al. 2001). In previous research on work-related ICT use during after-hours, Glavin and Schieman (2012), for example, have applied the JD-R model to examine its relationships to work-to-family conflict, excessive work pressures, schedule control, job authority, and decision-making latitude (see Sect. 4.3). Furthermore, Day et al. (2010) used it to provide a theoretical framework which posits ICT as both a demand and a resource.

In conclusion, the JD-R model posits differential outcomes of job demands and job resources, particularly with regard to exhaustion and work engagement. Some assumptions from the JD-R model will be integrated in our final conceptual framework. They will be discussed in Chap. 5.

2.7   Challenge-Hindrance Framework

The *challenge-hindrance framework* (Cavanaugh et al. 2000; LePine et al. 2005) highlights the distinction between two forms of self-reported work stressors: stressors that primarily hinder individuals and stressors that challenge individuals. The former, called *hindrance stressors* are likely to be appraised as hindering personal development and work-related accomplishment. Examples include red tape, organizational politics, role ambiguity, role conflict, hassles, or job insecurity. Thus, the work-related ICT use during non-work time may be considered as a hindrance stressor when it leads, for example, to work-life conflict. *Challenge stressors*, on the other hand, are likely to be appraised as both stressful and as promoting employee personal growth and achievement. Examples include time pressure, job complexity, high workload, job scope, and high levels of responsibility. According to this differentiation, supplemental work using ICT during after-hours may be perceived as a challenge stressor by increasing the overall daily workload. This framework explains why some job demands or other work circumstances, besides adverse effects, also have favorable effects on motivation and performance at the same time. Specifically, the assumption is that challenge stressors (as opposed to hindrance stressors) positively influence motivation and performance because they are accompanied by feelings of accomplishment and fulfillment (Cavanaugh et al. 2000).

Meta-analytic evidence supports the distinction between challenge stressors and hindrance stressors (Podsakoff et al. 2007; LePine et al. 2005). Although both challenge stressors and hindrance stressors are generally associated with higher experienced strains, there are differential relationships with job attitudes, turnover, motivation, and performance. Whereas challenge stressors are associated with more beneficial job attitudes (job satisfaction and organizational commitment), higher motivation, higher performance, and lower turnover indicators (turnover intentions, job search, turnover, and withdrawal behavior), the opposite was true for hindrance stressors.

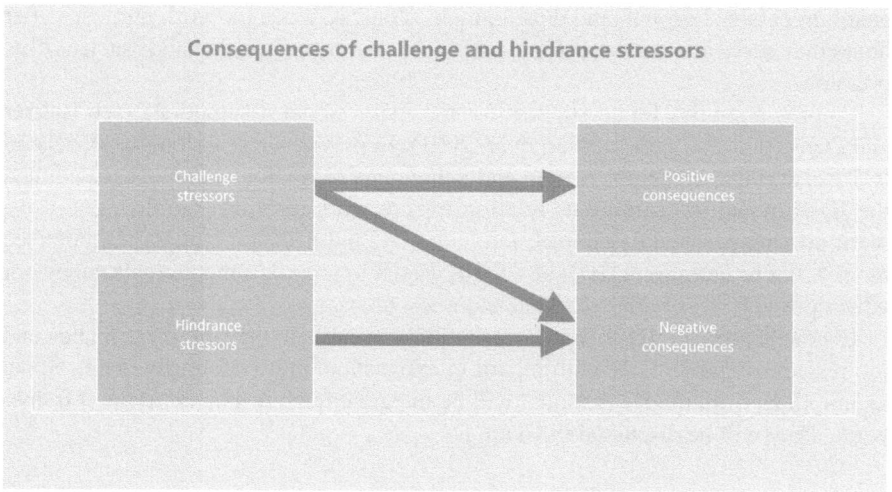

Fig. 2.6 The substantial assumptions of the challenge-hindrance framework

As far as we know, up to date, no study has yet examined the challenge-hindrance framework in the context of work-related ICT use during after-hours. Possible assumptions in our context according to the overall framework are: Work-related ICT use during non-work time is supposed to be increasing work-life conflict (the 'hindrance stressor hypothesis'). Otherwise, it should be supposed to be positively related to strain as well as to job satisfaction by increasing job scope (the 'challenge stressor hypothesis').

Taken together, challenge stressors might have more positive consequences than hindrance stressors, although both require substantial effort to deal with (as indicated in the higher level of strain). The empirical findings suggest that distinguishing among challenge- and hindrance stressors could be worthwhile for research as well as for practice (LePine et al. 2005). The differentiation between challenge stressors and hindrance stressors will be integrated, with minor amendments, in our final conceptual framework in Chap. 5. Thereby, the most important assumption for us is that challenge stressors may lead to both positive as well as negative outcomes, i.e., they contain a 'resourcing part' as well as a 'straining part', and hindrance stressors contain only a 'straining part' (see Fig. 2.6).

2.8   Action Theory

Action regulation theory (Hacker 1998, 2003; Frese and Zapf 1994) is a cognitive theory of work behavior. It assumes that all human behavior is organized towards goals. The action sequence can be described as consisting of multiple steps:

goal – orientation – planning – monitoring of execution – feedback. This action sequence is organized in a circle so that after feedback, the goal is activated again, and feedback is compared to the goal. If the goal is achieved, the action sequence ends. If the goal is not achieved, the sequence starts again. Actions, according to the theory, are regulated in a hierarchy from *unconscious to conscious regulation*. On the unconscious level, only sensory information is used to guide behavior; the regulation is free from consciousness and occurs fast and almost effortless. On the next level, the level of flexible action patterns, regulation can occur consciously but oftentimes one step in a sequence of steps triggers the next step, or behavior occurs as a result of external signals. Behavior is well-trained and comparable to habits or routines. Finally, on the level of intellectual regulation, the regulation is always conscious but slow.

This theory has important implications for the treatment of *stressors* at work (Sonnentag and Frese 2003). Stressors are defined as regulation problems as they disturb the regulation of actions (Frese and Zapf 1994). They can be classified according to their effect on the action sequence as regulation obstacles, regulation uncertainty, or overtaxing regulations. Regulation obstacles make action regulation more difficult. Regulation uncertainty refers to uncertainty about how to reach the goal. Overtaxing regulation means that the speed and intensity of action cannot be upheld for long. Examples for regulation obstacles include interruptions or organizational constraints (e.g., bad technological equipment), for regulation uncertainty lack of feedback, and for overtaxing regulation, time pressure or the requirement to concentrate. Moreover, action theory has a special theoretical framework that distinguishes between three work characteristics: regulation problems (stressors), regulation requirements (demands), and regulation possibilities (resources) (Frese and Zapf 1994). In the following text, we will use the generally established terms of stressors, demands, and resources. According to action theory, *demands* are "requirements necessary to do a particular task" (Zapf 1993, p. 86). An example variable is task complexity (low, e.g., in assembly line work). To do a complex task, it is necessary to solve many problems. This cannot be avoided. *Resources* are characterized as means to avoid, hinder, or deal with stressors (Frese 1989; Semmer 1990; Zapf and Semmer 2004). Zapf and Semmer (2004) differentiate between internal (e.g., self-efficacy and optimism) and external resources (e.g., social support and autonomy). The core resourcing variable is job control (comparable to decision latitude and autonomy).

The value of the three-way differentiation (stressors, demands, and/or resources) lies in the fact that it enables to propose *different relationships with outcomes*, such as recovery and well-being (Zapf and Semmer 2004). Thus, in general, *stressors* may impair recovery and well-being. For example, high overall workload caused by the supplemental work through ICT use during after-hours may lead to impairments of recovery and well-being. In contrast, *resources* may lead to improvements in recovery and well-being. In this context, ICT use can be considered as a resource, affecting well-being in several different ways. First, resources can affect recovery and well-being directly in a positive way. For example, the work-related ICT use during after-hours itself may lead to improvements of recovery and well-being.

Second, the resources affect recovery and well-being indirectly as they may reduce stressors, which, in turn, lose or at least decrease their negative effects on recovery and well-being. In our context, work-related ICT use during after-hours may reduce acute time pressure at work by enabling additional work at leisure time. As an escalation manager reported:

> Sometimes I get an e-mail from somebody that says, "What are you doing answering e-mail on the weekend?" It's one of my methods of stress management. I could work a normal 40-hour work week and then I'd be stressed out on the weekend and in the evenings. It's less stressful for me to put in the hours because I don't have a backlog building up. (Barley et al. 2011, p. 898)

Thirdly, resources may puffer the stressor-strain association. Even the awareness that work-related ICT may be used during after-hours may decrease the negative effect of perceived time pressure on well-being indicators because additional work is perceived as a way to cope with time pressure at work. Furthermore, the link between *demands* and recovery, or well-being, seems to be curvilinear. Thus, an optimal level on demands for every person is proposed (according to person-environment fit approach; Edwards et al. 1998; Edwards and Van Harrison 1993), and extreme low or high levels may be detrimental for employees. In our context, work-related ICT use during after-hours may facilitate well-being indicators for ambitious employees, but with the exception of extreme use (due for example to physiological factors, such as need for recovery). In sum, the consideration of three various work characteristics (stressors, demands, and resources) and their various associations to recovery and well-being does not emphasize the importance of job strain reduction, but especially the role of person-job fit (Zapf and Semmer 2004). Previous empirical studies have already confirmed the existence of the assumed different relationships (see, e.g., Dunckel 1985; Edwards et al. 1998; Zapf and Semmer 2004). To the best of our knowledge, no study has yet examined the antecedents or consequences of work-related ICT use during after-hours according to action theory.

Overall, we consider action theory as the basis for our conceptual work. Fundamental for our theoretical approach is the differentiation of the three work characteristics: stressors, demands, and resources. Equally important are the various, proposed links between stressors, demands, resources and our core concepts of recovery and well-being (e.g., linear negative, linear positive, buffering, and curvilinear effect). They will be subsequently reported in detail when developing the extended theoretical model below (see Chap. 5).

References

Ashforth, B. E., Kreiner, G. E., & Fugate, M. (2000). All in a day's work: Boundaries and micro role transitions. *Academy of Management Review, 25*(3), 472–491. doi:10.2307/259305.

Bakker, A. B., & Demerouti, E. (2007). The Job demands-resources model: State of the art. *Journal of Managerial Psychology, 22*(3), 309–328. doi:10.1108/02683940710733115.

Bakker, A. B., Demerouti, E., Taris, T. W., Schaufeli, W. B., & Schreurs, P. J. (2003). A multigroup analysis of the job demands-resources model in four home care organizations. *International Journal of Stress Management, 10*(1), 16–38. doi:10.1037/1072-5245.10.1.16.

Bandura, A. (1965). Vicarious processes: A case of no-trial learning. In L. Berkowitz (Ed.), *Advances in Experimental Social Psychology* (pp. 1–55). New York: Academic.

Bandura, A. (1977). *Social learning theory*. Englewood Cliffs: Prentice Hall.

Bandura, A. (1986). *Social foundations of thought and action: A social cognitive theory*. Englewood Cliffs: Prentice Hall.

Bandura, A., & Walters, R. H. (1963). *Social learning and personality development*. New York: Holt, Rinehart, Winston.

Barley, S. R., Meyerson, D. E., & Grodal, S. (2011). E-mail as a source and symbol of stress. *Organization Science, 22*(4), 887–906. doi:10.1287/orsc.1100.0573.

Battard, N., & Mangematin, V. (2013). Idiosyncratic distances: Impact of mobile technology practices on role segmentation and integration. *Technological Forecasting and Social Change, 80*(2), 231–242. doi:10.1016/j.techfore.2011.11.007.

Cavanaugh, M. A., Boswell, W. R., Roehling, M. V., & Boudreau, J. W. (2000). An empirical examination of self-reported work stress among U.S. managers. *Journal of Applied Psychology, 85*(1), 65–74. doi:10.1037/0021-9010.85.1.65.

Clark, S. C. (2000). Work/family border theory: A new theory of work/family balance. *Human Relations, 53*(6), 747–770. doi:10.1177/0018726700536001.

Collins, E. I., & Cox, A. L. (2014). *Out of work, out of mind? Smartphone use and work-life boundaries*. Socio-technical systems and Work-Home Boundaries Workshop. MobileHCI 2014, 2014.

Day, A., Scott, N., & Kelloway, E. K. (2010). Information and communication technology: Implications for job stress and employee well-being. In P. L. Perrewe & D. C. Ganster (Eds.), *New developments in theoretical and conceptual approaches to job stress* (Research in occupational stress and well being, Vol. 8, pp. 317–350). Bingley: Emerald.

Deci, E. L., & Ryan, R. M. (2000). The "what" and "why" of goal pursuits: human needs and the self-determination of behavior. *Psychological Inquiry, 11*(4), 227–268. doi:10.1207/S15327965PLI1104_01.

Demerouti, E., Bakker, A. B., Nachreiner, F., & Schaufeli, W. B. (2001). The job demands-resources model of burnout. *Journal of Applied Psychology, 86*(3), 499–512. doi:10.1037/0021-9010.86.3.499.

Dunckel, H. (1985). *Mehrfachbelastungen am Arbeitsplatz und psychosoziale Gesundheit: Psychologische Überlegungen und aktuarische Analysen*. Frankfurt am Main/New York: P. Lang.

Edwards, J. R., & Van Harrison, R. (1993). Job demands and worker health: Three-dimensional reexamination of the relationship between person-environment fit and strain. *Journal of Applied Psychology, 78*(4), 628–648.

Edwards, J. R., Caplan, R. D., & Harrison, R. V. (1998). Person-environment fit theory: Conceptual foundations, empirical evidence, and directions for future research. In C. L. Cooper (Ed.), *Theories of organizational stress* (pp. 28–67). Oxford/New York: Oxford University Press.

Fischer, T., & Riedl, R. (2015). Technostress in organizations: A cybernetic approach. *Proceedings der 12. Internationalen Tagung Wirtschaftsinformatik (WI 2015)*, 1453–1467.

Frese, M. (1989). Theoretical models of control and health. In S. L. Sauter, J. J. Hurrell, & C. L. Cooper (Eds.), *Job control and worker health* (pp. 108–128). Chichester/New York: Wiley.

Frese, M., & Zapf, D. (1994). Action as the core of work psychology: A German approach. In M. D. Dunnette, L. M. Hough, & H. C. Triandis (Eds.), *Handbook of industrial and organizational psychology* (Vol. 4, pp. 271–340). Palo Alto: Consulting Psychologists Press.

Gagné, M., & Deci, E. L. (2005). Self-determination theory and work motivation. *Journal of Organizational Behavior, 26*(4), 331–362. doi:10.1002/job.322.

Gagné, M., Forest, J., Vansteenkiste, M., Crevier-Braud, L., van den Broeck, A., Aspeli, A. K., et al. (2015). The multidimensional work motivation scale: Validation evidence in seven languages and nine countries. *European Journal of Work and Organizational Psychology, 24*(2), 178–196. doi:10.1080/1359432X.2013.877892.

Glavin, P., & Schieman, S. (2012). Work-family role blurring and work-family conflict: The moderating influence of job resources and job demands. *Work and Occupations, 39*(1), 71–98. doi:10.1177/0730888411406295.

Gorgievski, M., & Hobfoll, S. E. (2008). Work can burn us out or fire us up: Conservation of resources in burnout and engagement. In J. R. B. Halbesleben (Ed.), *Handbook of stress and burnout in health care* (pp. 7–22). New York: Nova.

Hacker, W. (1998). *Allgemeine Arbeitspsychologie: Psychische Regulation von Arbeitstätigkeiten.* Bern: H. Huber.

Hacker, W. (2003). Action regulation theory: A practical tool for the design of modern work processes? *European Journal of Work and Organizational Psychology, 12*(2), 105–130. doi:10.1080/13594320344000075.

Hobfoll, S. E. (1989). Conservation of resources: A new attempt at conceptualizing stress. *American Psychologist, 44*(3), 513–524. doi:10.1037/0003-066X.44.3.513.

Hobfoll, S. E. (2011). Conservation of resource caravans and engaged settings. *Journal of Occupational and Organizational Psychology, 84*(1), 116–122. doi:10.1111/j.2044-8325.2010.02016.x.

Hobfoll, S. E., & Freedy, J. (1993). Conservation of resources. In W. B. Schaufeli (Ed.), *Professional burnout: Recent developments in theory and research* (pp. 115–133). Washington, DC: Taylor & Francis.

Lazarus, R. S. (2006). *Stress and emotion: A new synthesis.* New York: Springer.

Lazarus, R. S., & Folkman, S. (1984). *Stress, appraisal, and coping.* New York: Springer.

Lei, C. F., & Ngai, E. W. T. (2014). *The double-edged nature of technostress on work performance: A research model and research agenda: Completed research paper.* Thirty fifth international conference on information systems, Auckland.

LePine, J. A., Podsakoff, N. P., & LePine, M. A. (2005). A meta-analytic test of the challenge stressor-hindrance stressor framework: An explanation for inconsistent relationships among stressors and performance. *The Academy of Management Journal, 48*(5), 764–775.

MacCormick, J. S., Dery, K., & Kolb, D. G. (2012). Engaged or just connected? Smartphones and employee engagement. *Organizational Dynamics, 41*(3), 194–201. doi:10.1016/j.orgdyn.2012.03.007.

Mazmanian, M. A., Orlikowski, W. J., & Yates, J. (2013). The autonomy paradox: The implications of mobile email devices for knowledge professionals. *Organization Science, 24*(5), 1337–1357. doi:10.1287/orsc.1120.0806.

Meijman, T. F., & Mulder, G. (1998). Psychological aspects of workload. In P. J. D. Drenth & H. Thierry (Eds.), *Handbook of work and organizational psychology* (pp. 5–33). Hove: Psychology Press.

Nippert-Eng, C. E. (1996). *Home and work: Negotiating boundaries through everyday life.* Chicago: University of Chicago Press.

Ohly, S., & Latour, A. (2014). Work-related smartphone use and well-being in the evening: The role of autonomous and controlled motivation. *Journal of Personnel Psychology, 13*(4), 174–183. doi:10.1027/1866-5888/a000114.

Olson-Buchanan, J. B., & Boswell, W. R. (2006). Blurring boundaries: Correlates of integration and segmentation between work and nonwork. *Journal of Vocational Behavior, 68*(3), 432–445. doi:10.1016/j.jvb.2005.10.006.

Park, Y., & Jex, S. M. (2011). Work-home boundary management using communication and information technology. *International Journal of Stress Management, 18*(2), 133–152. doi:10.1037/a0022759.

Park, Y., Fritz, C., & Jex, S. M. (2011). Relationships between work-home segmentation and psychological detachment from work: The role of communication technology use at home. *Journal of Occupational Health Psychology, 16*(4), 457–467. doi:10.1037/a0023594.

Podsakoff, N. P., LePine, J. A., & LePine, M. A. (2007). Differential challenge stressor-hindrance stressor relationships with job attitudes, turnover intentions, turnover, and withdrawal behavior: a meta-analysis. *Journal of Applied Psychology, 92*(2), 438–454. doi:10.1037/0021-9010.92.2.438.

Richardson, K. M., & Thompson, C. A. (2012). High tech tethers and work-family conflict: A conservation of resources approach. *Engineering Management Research, 1*(1), 29–43. doi:10.5539/emr.v1n1p29.

Sayah, S. (2013). Managing work-life boundaries with information and communication technologies: the case of independent contractors. *New Technology, Work and Employment, 28*(3), 179–196. doi:10.1111/ntwe.12016.

Semmer, N. K. (1990). Stress und Kontrollverlust. In F. Frei & I. Udris (Eds.), *Das Bild der Arbeit* (pp. 190–207). Bern: H. Huber.

Sonnentag, S., & Frese, M. (2003). Stress in organizations. In W. C. Borman, D. R. Ilgen, & R. Klimoski (Eds.), *Handbook of psychology* (Industrial and Organizational psychology, Vol. 12, pp. 453–491). New York/Chichester: Wiley.

Tremblay, M. A., Blanchard, C. M., Taylor, S., Pelletier, L. G., & Villeneuve, M. (2009). Work extrinsic and intrinsic motivation scale: Its value for organizational psychology research. *Canadian Journal of Behavioural Science/Revue canadienne des Sciences du comportement, 41*(4), 213–226. doi:10.1037/a0015167.

Venkatesh, V., Morris, M. G., Davis, G. B., & Davis, F. D. (2003). User acceptance of information technology: Toward a unified view. *Management Information Systems Quarterly, 27*(3).

Ward, S., & Steptoe-Warren, G. (2013). A conservation of resources approach to BlackBerry use, work-family conflict and well-being: Job control and psychological detachment from work as potential mediators. *Engineering Management Research, 3*(1), 8–23. doi:10.5539/emr.v3n1p8.

Yin, P., Davison, R. M., Bian, Y., Wu, J., & Liang, L. (2014). The sources and consequences of mobile technostress in the workplace. Paper 144. *PACIS 2014 Proceedings*.

Zapf, D. (1993). Stress-oriented analysis of computerized office work. *European Work and Organizational Psychologist, 3*(2), 85–100. doi:10.1080/09602009308408580.

Zapf, D., & Semmer, N. K. (2004). Stress und Gesundheit in Organisationen. In H. Schuler (Ed.), *Enzyklopädie der Psychologie: Organisationspsychologie – Grundlagen und Personalpsychologie* (Wirtschafts- Organisations- und Arbeitspsychologie, Vol. 3, pp. 1007–1112). Göttingen: Hogrefe.

Chapter 3
Conceptualization of Core Concepts

In the present work, we focus on individuals' recovery during non-work time and well-being as two important concepts in work and industrial psychology. In this chapter, we present the conceptualizing of these core concepts. First, we describe theories of work recovery and offer an overview of its operationalization in empirical research. Second, we introduce our conceptualization of recovery. Finally, we look at theoretical frameworks of subjective well-being, choose a concept that best suits our purposes and give an overview about the operationalization of well-being in some selected studies.

3.1 Conceptualization of Employee Recovery

In this section, we focus on individuals' recovery during non-work time as an important concept in the context of employees' well-being (e.g., Newman et al. 2014). First, we give an established definition of recovery. Then, we shortly describe three common recovery theories. Finally, we describe some empirical studies, which adopted various sorts of recovery operationalization and present our conceptualization of recovery.

3.1.1 Theories of Recovery

In this work, we use the common definition of *recovery* by Geurts and Sonnentag (2006), namely, „a process of psychophysiological unwinding after effort expenditure"(p. 482). In this section, we shortly describe three common recovery theories. First, according to the *effort-recovery theory* (Meijman and Mulder 1998), employees need sufficient recovery from work-related effort; otherwise the acute physiological, behavioral, and subjective load reactions (e.g., elevated blood pressure and fatigue) caused by the effort expenditure in the job may have serious implications for their health and well-being. Geurts and Sonnentag (2006) argue that the extended exposure to work effort after hours hinders recovery by prolonging the same psychophysiological activation as on the job. To complete the recovery

© The Author(s) 2016
L. Ďuranová, S. Ohly, *Persistent Work-Related Technology Use, Recovery and Well-being Processes*, SpringerBriefs in Psychology,
DOI 10.1007/978-3-319-24759-5_3

process after hours, the authors suggest that employees should not engage in activities that appeal to the same activation systems that were already loaded during work, or they should avoid completely all activities that require effort. Picturing a job of some knowledge worker who experiences high demands on her/his cognitive resources (e.g., ability to concentrate) at work, in regard to recovery, she/he should not practice non-work activities which demand the same sort of effort, such as computer use, whereas physical activities, such as jogging or no-effort activities, such as watching TV would less hinder or even facilitate the recovery process. Meijman and Mulder (1998) assume that during the optimal recovery process (e.g., by taking a break) the acute load reactions return to baseline levels. Furthermore, this recovery process is an important prerequisite for well-being and for adequate performance on the next workday. Otherwise, not recovered employees have to compensate their suboptimal state through an additional effort. The chronic sustaining of physiological activation and long-term incomplete recovery may lead to chronic health impairment (Geurts and Sonnentag 2006). To conclude, according to effort-recovery theory, the non-work activities that draw on the same psychophysiological activation or that use the same resources as during work impede the recovery process, as an executive expressed with regard to the work-related use of ICT during after-hours:

> ... there are no more complete breaks. It is a sort of permanent connection that did not exist before." "Previously, when I caught transport to and from work, I would've read the paper, read a book ..." "You just can't get away from it. (MacCormick et al. 2012, emphasis in original)

As Sonnentag and Natter (2004) state, the effort-recovery model and the *conservation of resources* theory (COR; Hobfoll 1989, 2011) complement each other. COR theory suggests that people strive to maintain, defend, and build resources they value, and they experience psychological stress should these resources be threatened, lost, or not gained after investing in them (Hobfoll 1989). Therefore, during leisure time, employees strive to restore their resources depleted at work. Because the COR theory was extensively introduced in Sect. 2.4 above, at this point we focus only on the difference to the effort-recovery theory. Hence, the additional value of COR theory in the context of recovery is the assumption that available resources may help to restore the threatened resources. Therefore, resources can be restored not only by resting but also by investing other resources.

A further model appropriate to explain recovery, respectively need for recovery, is the *self-control model* (Baumeister et al. 1998; Muraven and Baumeister 2000) which suggests that continuous acts of self-control draw from a limited pool of resources. This resource pool is vulnerable to depletion. In other words, the ability to control one's impulses or, in more general terms, to self-regulate, is limited. Once self-control is required by a task (e.g., to control one's emotions or to resist the temptation to eat a forbidden food) the capacity to control oneself in a subsequent task is reduced. This depletion of self-control (called *ego depletion*) is experienced by the person as fatigue (Hagger et al. 2010). Once depleted, people find subsequent work activities more demanding and become vulnerable to distractions and impulses (Baumeister and Vohs 2007). Ego depletion can be reversed by the replenishment of resources through glucose intake or by rest and relaxation.

The assumptions of the presented theories will be integrated in the theoretical framework below.

3.1.2 Operationalization of Recovery

The aim of this section is to provide several ways capturing recovery. Thus, in the next paragraphs, we consider some empirical studies that use various concepts as indicators of recovery. We choose them in regard to their variability in the conceptualizing of recovery. For example, Bakker et al. (2013) asked participants directly to assess their *level of recovery* with focus on a specific period, even before bedtime. An example item is "Right now, I feel recovered from work" (p. 92). *Fatigue* is conceptually closely related to recovery. Fatigue refers to a state of high negative and low positive affect with low arousal. The subjective experience can be described by the adjectives such as: "sleepy", "tired", "sluggish", and "drowsy" (Watson and Clark 1994). Thus, fatigue may be seen as an indicator of an unsuccessful recovery process or as a consequence of failed recovery (Demerouti et al. 2009). *State-depletion* is similar to fatigue. It refers to a feeling of not being able to exert self-control, and is subjectively experienced as fatigue (Hagger et al. 2010). In the study by Lanaj et al. (2014), state depletion was measured in the morning with the following sample items: "I feel drained" and "Right now, it would take a lot of effort for me to concentrate on something" (p. 15). After fatigue, *need for recovery* follows. According to Sonnentag and Zijlstra (2006), need for recovery means "a sense of urgency to take a break from the demands" (p. 331). They measured it, among others, by the following item: "Today I would have needed more time for relaxing and recovering from work" (p. 335).

Furthermore, there are some *psychological mechanisms in the leisure time* relevant to recovery. They are the experiences of psychological detachment (see, e.g., Barber and Jenkins 2014; Derks and Bakker 2014; Newman et al. 2014; Siltaloppi et al. 2009; Sonnentag et al. 2008; Sonnentag and Fritz 2007), relaxation (see, e.g., Newman et al. 2014; Siltaloppi et al. 2009; Sonnentag et al. 2008; Sonnentag and Fritz 2007), mastery (see, e.g., Newman et al. 2014; Siltaloppi et al. 2009; Sonnentag et al. 2008; Sonnentag and Fritz 2007), control (see, e.g., Newman et al. 2014; Siltaloppi et al. 2009; Sonnentag and Fritz 2007), meaning (see, e.g., Newman et al. 2014) and affiliation (see, e.g., Newman et al. 2014). The most investigated concept in this context is psychological detachment. Psychological detachment refers to a state of mind in which employees feel mentally disconnected from the work situation (Etzion et al. 1998; Sonnentag 2012) and it is considered as a "core recovery experience" (Sonnentag and Fritz 2015). Employees who are detached do not think about work-related issues after hours. To assess psychological detachment, items such as "I forget about work" and "I distance myself from my work" (Sonnentag and Fritz 2007, p. 213) are used. Previous research has already shown some associations between psychological detachment and indicators of well-being, such as life satisfaction and work engagement (for a review see Sonnentag and Fritz 2015).

Another way to operationalize recovery, specifically recovery in the morning, is to focus on *sleep*. Research focusing on sleep has examined a broad range of its indicators. One way to do this was shown by Sonnentag et al. (2008) as they assessed the sleep quality. The single item derived from the Pittsburgh Sleep Quality Index (Buysse et al. 1989) was: "How do you evaluate this night's sleep?" (p. 677). Gustafsson et al. (2008) operationalized recovery through several indicators, such as morning recovery, fatigue, or sleep problems. An example item was "During the past week, have you had difficulties sleeping (difficulties falling asleep, waking too early due to work) because work-related thoughts have kept you awake?" and "How well do you normally sleep?" (p. 26). Another option to measure sleep sufficiency is to ask participants to refer to the amount of hours they slept. For instance, Lanaj et al. (2014) measured sleep quantity with this single item adapted from the Pittsburgh Sleep Quality Index (Buysse et al. 1989): "How many hours of actual sleep did you get last night (this may be different than the number of hours you spent in bed)?" (p. 15). The third mentioned option to assess sleep seems to be the most objective. On the other hand, it can be considered as an antecedent of the other sleep concepts.

A further step towards capturing recovery more objectively through *physiological markers*, is, for example, to operationalize it with levels of cortisol (e.g., Gustafsson et al. 2008; Saxbe et al. 2008), catecholamine (for a review see Sonnentag and Fritz 2006), and other indicators of allostatic load (e.g., von Thiele et al. 2006). In a study among white-collar workers, Gustafsson et al. (2008) investigated how recovery self-ratings relate to cortisol levels. Participants sampled saliva every second hour across two working days. The results showed that poor recovery was associated in particular with high levels of morning cortisol. Therefore, the authors concluded that self-ratings of recovery provide valid information on physiological recovery in terms of cortisol level, particularly in the morning.

In summary, research delivers various operationalization types of recovery. They differentiate in their level of directness in measuring recovery from direct (such as question for recovery level) to rather non-direct (such as capturing of non-work events).

From this short overview, it is apparent that there are several operationalizations of recovery. In the following chapters, we will consider all the variants of the recovery concept mentioned above.

3.2 Conceptualization of Employee Well-Being

In the present work, we focus on employee well-being for three reasons. First, establishing employee well-being is an aim in itself (see, e.g., Hacker 1998). Second, poor health is associated with costs for individuals, organizations and societies (see, e.g., Sonnentag and Frese 2003). Third, according to 'the happy–productive worker hypothesis', psychological well-being is associated with positive performance outcomes which benefit organizations (Wright and Cropanzano 2000, p. 84).

3.2.1 Theories of Well-Being

There are several well-known theories of well-being (e.g., Diener et al. 1999; Ryff 1989; Warr 1994). Most of the developed models up to now are multidimensional. For example, Ryff (1989) derives from literature the following aspects of well-being: self-acceptance, positive relations with others, autonomy, environmental mastery, purpose in life, and personal growth. The model by Warr (1994) consists of these four well-being components: affective well-being, aspiration, competence, and autonomy. Van Horn et al. (2004) developed a model for occupational well-being composed of affective, professional, social, cognitive, and psychosomatic well-being. Thereby, they have built on the models by Ryff (1989) and Warr (1994). Concerning the purposes of our work, none of these models seems to be sufficiently comprehensive as they consider in each case only certain aspects of well-being. The search for a broader conceptualizing of well-being has led us to work by Diener et al. (1999). Indeed, they described subjective well-being (SWB) "as a general area of scientific interest rather than a single specific construct" (p. 277). Accordingly, they present four major divisions inclusive subdivisions of the concept. According to Diener et al. (1999), SWB may be conceptualized as *life satisfaction* (desire to change life, satisfaction with current life, satisfaction with past, satisfaction with future, and satisfaction with others' views of one's life), *domain satisfaction* (work, family, leisure, health, finances, self, and one's group), *pleasant affect* (joy, elation, contentment, pride, affection, happiness, and ecstasy), and *unpleasant affect* (guilt and shame, sadness, anxiety and worry, anger, stress, depression, and envy). In our opinion, this conceptualizing best covers all substantial topics regarding the possible consequences of work-related ICT use during after-hours. Particularly, it refers to a broad term of SWB that means a wide range of moods, emotions, general- and domain-specific satisfaction. As a consequence, in the next sections, we will consequently use the conceptualization of SWB by Diener et al. (1999).

3.2.2 Operationalization of Well-Being

As regards the conceptual comparison with the previous research, we next present some examples of studies, which adopted various sorts of well-being operationalization. We choose them in regard to their variability in the conceptualizing of well-being. Thus, there are several ways to conceptualize well-being: For example, Matthews et al. (2014) used a *general concept* of SWB. A sample item is "Thinking about the past 30 days, how often have you been able to enjoy your normal day to day activities?" (p. 173). Furthermore, Magsamen-Conrad et al. (2014) assessed well-being with the Warwick-Edinburgh Mental Well-Being Scale (Tennant et al. 2007) that considers various dimensions of SWB. An example item is "I've been feeling optimistic about the future" (p. 27). Exclusively *life satisfaction* was focused, for example, by Hahn and Dormann (2013). Likewise, Widmer et al. (2012) captured positive attitude towards life with items such as: "My future looks good" and

"I cope well with the things in my life that can't be changed". In regard to *domain satisfactions* and *affects*, for example, Siltaloppi et al. (2009) focused on occupational well-being and conceptualized it as need for recovery, job exhaustion, and work engagement. Also Schmitt et al. (2012) concentrated on the occupational well-being which was measured through job satisfaction and fatigue. Grebner et al. (2005) measured SWB in terms of general or context-free well-being (psychosomatic complaints and irritated reactions), job-related well-being (job satisfaction), and spillover from work into non-work domains (rumination about work problems and inability to switch off job-related matters). Sonnentag and Fritz (2015) considered in their review positive and negative affective states as well as life satisfaction and work engagement as indicators of SWB to examine short-term reactions and long-term reactions of psychological detachment. Bakker et al. (2013) measured following indicators of daily well-being in the evening: evening happiness, momentary vigor before bedtime, and momentary recovery before bedtime. Sonnentag and Natter (2004) assessed also a well-being-state with the constructs of vigor, depression, and fatigue. However, the last named studies considered vigor as an indicator of well-being, although it is actually controversial in research. For example, Ryan and Frederick (1997) argue that vigor is not *always* an indicator of SWB because there are many positive affective states in which individuals do not experience having a lot of energy (e.g., serenity), and, on the other hand, there are other negative states that are characterized by high amount of energy, and are rather not indicators of SWB (e.g., anger). Thus, we would consider vigor as an energy resource (Hobfoll 1989) and, therefore, as a potential indicator of well-being but only in the absence of negative states.

In summary, the concept of SWB is multidimensional. Therefore, it can be operationalized with diverse constructs. Some of them are rather stable (like affective organizational commitment) and some of them are constantly changing (like affects). Indeed, in previous empirical research SWB has been operationalized in multiple ways. Thereby, operationalizations include both positive and negative aspects of SWB. Furthermore, our short overview shows that it is not very common that SWB is measured by all the constructs as the concept by Diener et al. (1999) provides. More common is the focus on a few specific aspects of SWB. Furthermore, at this point it should be noted that most researchers do not justify their choice of indicators of well-being. The theoretical foundations are often lacking in the empirical research on SWB.

In this work, we follow the theoretical concept by Diener et al. (1999) and distinguish four components of SWB: life satisfaction, domain satisfaction, pleasant affect, and unpleasant affect. The examples of constructs belonging to SWB can be seen in Table 3.1. At this point, it should be noted that in the following text we label the components of work/organization satisfaction by Diener et al. (1999) also by the term *organizational well-being*. We conceptualize organizational well-being as all the attitudes toward an organization (like affective commitment or organizational identification) and work (like work engagement and job involvement) which are highly desirable by organizations as they are supposed to contribute to important outcomes, such as individual performance and overall organizational performance.

Table 3.1 Components of subjective well-being following Diener et al. (1999)

Life satisfaction	Domain satisfactions	Pleasant/ positive affect	Unpleasant/ negative affect
Desire to change life	Work/Organization	Joy	Guilt
Satisfaction with current life (Hahn and Dormann 2013; Widmer et al. 2012)	Job satisfaction (Grebner et al. 2005; Schmitt et al. 2012)	Elation	Shame
Satisfaction with past (Matthews et al. 2014)	Work engagement (Siltaloppi et al. 2009; Sonnentag and Fritz 2014)	Contentment	Sadness
Satisfaction with future (Magsamen-Conrad et al. 2014; Widmer et al. 2012)	Family	Pride	Anxiety
Satisfaction with others' views of one's life	Rumination about work problems at home (Grebner et al. 2005)	Affection	Worry
	Inability to switch off job-related matters (Grebner et al. 2005)	Happiness (Bakker et al. 2013)	Anger
	Leisure	Ecstasy	Stress
	Health	Vigor (Bakker et al. 2013; Sonnentag and Natter 2004)	Job exhaustion (Siltaloppi et al. 2009)
	Psychosomatic complaints and irritated reactions (Grebner et al. 2005)	Recovery (Bakker et al. 2013)	Depression (Sonnentag and Natter 2004)
	Finances		Envy
	Self		Need for recovery (Siltaloppi et al. 2009)
	One's group		Fatigue (Schmitt et al. 2012; Sonnentag and Natter 2004)

Note: We listed references from the examples of empirical studies noted above which captured similar concepts to Diener et al. (1999). Additional concepts which are not explicitly mentioned by Diener et al. (1999) are given in italics

References

Bakker, A. B., Demerouti, E., Oerlemans, W., & Sonnentag, S. (2013). Workaholism and daily recovery: A day reconstruction study of leisure activities. *Journal of Organizational Behavior, 34*(1), 87–107. doi:10.1002/job.1796.

Barber, L. K., & Jenkins, J. S. (2014). Creating technological boundaries to protect bedtime: Examining work-home boundary management, psychological detachment and sleep. *Stress and Health, 30*(3), 259–264. doi:10.1002/smi.2536.

Baumeister, R. F., & Vohs, K. D. (2007). Self-regulation, ego depletion, and motivation. *Social and Personality Psychology Compass, 1*(1), 115–128. doi:10.1111/j.1751-9004.2007.00001.x.

Baumeister, R. F., Bratslavsky, E., Muraven, M., & Tice, D. M. (1998). Ego depletion: Is the active self a limited resource? *Journal of Personality and Social Psychology, 74*(5), 1252–1265. doi:10.1037/0022-3514.74.5.1252.

Buysse, D. J., Reynolds, C. F., Monk, T. H., Berman, S. R., & Kupfer, D. J. (1989). The Pittsburgh sleep quality index: A new instrument for psychiatric practice and research. *Psychiatry Research, 28*(2), 193–213. doi:10.1016/0165-1781(89)90047-4.

Demerouti, E., Bakker, A. B., Geurts, S. A., & Taris, T. W. (2009). Daily recovery from work-related effort during non-work time. In S. Sonnentag, P. L. Perrewe, & D. C. Ganster (Eds.), *Current perspectives on job-stress recovery* (Research in occupational stress and well-being, Vol. 7). Bingley: Emerald/JAI Press.

Derks, D., & Bakker, A. B. (2014). Smartphone use, work-home interference, and burnout: A diary study on the role of recovery. *Applied Psychology, 63*(3), 411–440. doi:10.1111/j.1464-0597.2012.00530.x.

Diener, E., Suh, E. M., Lucas, R. E., & Smith, H. L. (1999). Subjective well-being: Three decades of progress. *Psychological Bulletin, 125*(2), 276–302. doi:10.1037//0033-2909.125.2.276.

Etzion, D., Eden, D., & Lapidot, Y. (1998). Relief from job stressors and burnout: Reserve service as a respite. *Journal of Applied Psychology, 83*(4), 577–585. doi:10.1037/0021-9010.83.4.577.

Geurts, S. A. E., & Sonnentag, S. (2006). Recovery as an explanatory mechanism in the relation between acute stress reactions and chronic health impairment. *Scandinavian Journal of Work, Environment & Health, 32*(6), 482–492. doi:10.5271/sjweh.1053.

Grebner, S., Semmer, N. K., & Elfering, A. (2005). Working conditions and three types of well-being: A longitudinal study with self-report and rating data. *Journal of Occupational Health Psychology, 10*(1), 31–43. doi:10.1037/1076-8998.10.1.31.

Gustafsson, K., Lindfors, P., Aronsson, G., & Lundberg, U. (2008). Relationships between self-rating of recovery from work and morning salivary cortisol. *Journal of Occupational Health, 50*(1), 24–30. doi:10.1539/joh.50.24.

Hacker, W. (1998). *Allgemeine Arbeitspsychologie: Psychische Regulation von Arbeitstätigkeiten.* Bern: H. Huber.

Hagger, M. S., Wood, C., Stiff, C., & Chatzisarantis, N. L. D. (2010). Ego depletion and the strength model of self-control: A meta-analysis. *Psychological Bulletin, 136*(4), 495–525. doi:10.1037/a0019486.

Hahn, V. C., & Dormann, C. (2013). The role of partners and children for employees' psychological detachment from work and well-being. *Journal of Applied Psychology, 98*(1), 26–36. doi:10.1037/a0030650.

Hobfoll, S. E. (1989). Conservation of resources: A new attempt at conceptualizing stress. *American Psychologist, 44*(3), 513–524. doi:10.1037/0003-066X.44.3.513.

Hobfoll, S. E. (2011). Conservation of resource caravans and engaged settings. *Journal of Occupational and Organizational Psychology, 84*(1), 116–122. doi:10.1111/j.2044-8325.2010.02016.x.

Lanaj, K., Johnson, R. E., & Barnes, C. M. (2014). Beginning the workday yet already depleted? Consequences of late-night smartphone use and sleep. *Organizational Behavior and Human Decision Processes, 124*(1), 11–23. doi:10.1016/j.obhdp.2014.01.001.

MacCormick, J. S., Dery, K., & Kolb, D. G. (2012). Engaged or just connected? Smartphones and employee engagement. *Organizational Dynamics, 41*(3), 194–201. doi:10.1016/j.orgdyn.2012.03.007.

Magsamen-Conrad, K., Billotte-Verhoff, C., & Greene, K. (2014). Technology addiction's contribution to mental wellbeing: The positive effect of online social capital. *Computers in Human Behavior, 40,* 23–30. doi:10.1016/j.chb.2014.07.014.

Matthews, R. A., Mills, M. J., Trout, R. C., & English, L. (2014). Family-supportive supervisor behaviors, work engagement, and subjective well-being: a contextually dependent mediated process. *Journal of Occupational Health Psychology, 19*(2), 168–181. doi:10.1037/a0036012.

Meijman, T. F., & Mulder, G. (1998). Psychological aspects of workload. In P. J. D. Drenth & H. Thierry (Eds.), *Handbook of work and organizational psychology* (pp. 5–33). Hove: Psychology Press.

Muraven, M., & Baumeister, R. F. (2000). Self-regulation and depletion of limited resources: Does self-control resemble a muscle? *Psychological Bulletin, 126*(2), 247–259. doi:10.1037/0033-2909.126.2.247.

Newman, D. B., Tay, L., & Diener, E. (2014). Leisure and subjective well-being: A model of psychological mechanisms as mediating factors. *Journal of Happiness Studies, 15*(3), 555–578. doi:10.1007/s10902-013-9435-x.

Ryan, R. M., & Frederick, C. (1997). On energy, personality, and health: Subjective vitality as a dynamic reflection of well-being. *Journal of Personality, 65*(3), 529–565. doi:10.1111/j.1467-6494.1997.tb00326.x.

Ryff, C. D. (1989). Happiness is everything, or is it? Explorations on the meaning of psychological well-being. *Journal of Personality and Social Psychology, 57*(6), 1069–1081. doi:10.1037/0022-3514.57.6.1069.

Saxbe, D. E., Repetti, R. L., & Nishina, A. (2008). Marital satisfaction, recovery from work, and diurnal cortisol among men and women. *Health psychol, 27*(1), 15–25. doi:10.1037/0278-6133.27.1.15.

Schmitt, A., Zacher, H., & Frese, M. (2012). The buffering effect of selection, optimization, and compensation strategy use on the relationship between problem solving demands and occupational well-being: A daily diary study. *Journal of Occupational Health Psychology, 17*(2), 139–149. doi:10.1037/a0027054.

Siltaloppi, M., Kinnunen, U., & Feldt, T. (2009). Recovery experiences as moderators between psychosocial work characteristics and occupational well-being. *Work & Stress, 23*(4), 330–348. doi:10.1080/02678370903415572.

Sonnentag, S. (2012). Psychological detachment from work during leisure time: The benefits of mentally disengaging from work. *Current Directions in Psychological Science, 21*(2), 114–118. doi:10.1177/0963721411434979.

Sonnentag, S., & Frese, M. (2003). Stress in organizations. In W. C. Borman, D. R. Ilgen, & R. Klimoski (Eds.), *Handbook of psychology* (Industrial and organizational psychology, Vol. 12). New York/Chichester: Wiley.

Sonnentag, S., & Fritz, C. (2006). Endocrinological processes associated with job stress: Catecholamine and cortisol responses to acute and chronic stressors. In P. L. Perrewe & D. C. Ganster (Eds.), *Employee health, coping and methodologies* (Research in occupational stress and well being, Vol. 5, pp. 1–59). Amsterdam/Boston: Elsevier.

Sonnentag, S., & Fritz, C. (2007). The recovery experience questionnaire: Development and validation of a measure for assessing recuperation and unwinding from work. *Journal of Occupational Health Psychology, 12*(3), 204–221. doi:10.1037/1076-8998.12.3.204.

Sonnentag, S., & Fritz, C. (2015). Recovery from job stress: The stressor-detachment model as an integrative framework. *Journal of Organizational Behavior, 36*(S1), S72–S103. doi:10.1002/job.1924.

Sonnentag, S., & Natter, E. (2004). Flight attendants' daily recovery from work: Is there no place like home? *International Journal of Stress Management, 11*(4), 366–391. doi:10.1037/1072-5245.11.4.366.

Sonnentag, S., & Zijlstra, F. R. H. (2006). Job characteristics and off-job activities as predictors of need for recovery, well-being, and fatigue. *Journal of Applied Psychology, 91*(2), 330–350. doi:10.1037/0021-9010.91.2.330.

Sonnentag, S., Binnewies, C., & Mojza, E. J. (2008). "Did you have a nice evening?" A day-level study on recovery experiences, sleep, and affect. *Journal of Applied Psychology, 93*(3), 674–684. doi:10.1037/0021-9010.93.3.674.

Tennant, R., Hiller, L., Fishwick, R., Platt, S., Joseph, S., Weich, S., et al. (2007). The Warwick-Edinburgh mental well-being scale (WEMWBS): Development and UK validation. *Health and Quality of Life Outcomes, 5*, 63. doi:10.1186/1477-7525-5-63.

Van Horn, J. E., Taris, T. W., Schaufeli, W. B., & Schreurs, P. J. (2004). The structure of occupational well-being: A study among Dutch teachers. *Journal of Occupational and Organizational Psychology, 77*(3), 365–375. doi:10.1348/0963179041752718.

von Thiele, U., Lindfors, P., & Lundberg, U. (2006). Self-rated recovery from work stress and allostatic load in women. *Journal of Psychosomatic Research, 61*(2), 237–242. doi:10.1016/j.jpsychores.2006.01.015.

Warr, P. (1994). A conceptual framework for the study of work and mental health. *Work & Stress, 8*(2), 84–97. doi:10.1080/02678379408259982.

Watson, D., & Clark, L. A. (1994). *The PANAS-X: Manual for the positive and negative affect schedule – Expanded form*. Ames: The University of Iowa.

Widmer, P. S., Semmer, N. K., Kälin, W., Jacobshagen, N., & Meier, L. L. (2012). The ambivalence of challenge stressors: Time pressure associated with both negative and positive well-being. *Journal of Vocational Behavior, 80*(2), 422–433. doi:10.1016/j.jvb.2011.09.006.

Wright, T. A., & Cropanzano, R. (2000). Psychological well-being and job satisfaction as predictors of job performance. *Journal of Occupational Health Psychology, 5*(1), 84–94. doi:10.1037/1076-8998.5.1.84.

Chapter 4
Empirical Findings

To date, only few empirical studies have examined antecedents and/or consequences of work-related ICT use during non-work time. In this chapter, we provide a tabular overview of these studies to demonstrate the current state of research in this field. At first, we offer an extensive overview of operationalization of ICT use (see Table 4.1). Then, we present empirical evidence and assumptions related to predictors and outcomes of work-related ICT use after hours (see Tables 4.2, 4.3, 4.4, and 4.5). The results will contribute to the foregoing theoretical basis of our final research model that will be described below. Our aim is to emphasize the importance of this research field viewing the previous empirical results of ICT use, and to show some research gaps especially with regard to individual's recovery and well-being.

For our purposes, we first conducted a literature search of journal articles published until 2014. Then, we reviewed the abstracts first and then the full texts. The articles had to be written in English and peer-reviewed. We included studies that presented empirical evidence on the antecedents and consequences of work-related ICT use during non-work time. Several exceptions are discussed in Sect. 4.1. Conceptual papers as well as qualitative studies were excluded. Furthermore, only studies focusing on the use of new technologies with internet access were taken into consideration. Finally, after examining all inclusion and exclusion criteria, 36 relevant articles remained. In total, 37 studies were identified from the 36 articles. They are presented in Table 4.1. Overall, empirical evidence seems to be rare up to now. Partially, it is due to our rather narrow inclusion criteria, especially focusing on research on work-related ICT use only as well as only during after-hours. Due to the scarcity of quantitative research on our main concept, we offer additionally a short overview of selected qualitative studies and conceptual papers in this research field (see Table 4.6). In doing this, we expect that they will broaden our mind when constructing the final theoretical framework.

In the following sections, we start with an overview of operationalization of ICT use. Then we describe several empirical results concerning predictors and outcomes of ICT use. Moreover, we present studies that consider ICT use as mediator or moderator in specific proposed relationships. Afterward, we consider a conceptual paper and selected qualitative studies on work-related ICT use during non-work time. Finally, we discuss the research gaps especially with regard to individuals' recovery and well-being.

© The Author(s) 2016
L. Ďuranová, S. Ohly, *Persistent Work-Related Technology Use, Recovery and Well-being Processes*, SpringerBriefs in Psychology,
DOI 10.1007/978-3-319-24759-5_4

4.1   Operationalization of Work-Related ICT Use during After-Hours

To begin with, a uniform understanding of the research topic is substantial. Previous research on work-related use of ICT during after-hours has termed it as 'work connectivity behavior during non-work time' (Richardson and Benbunan-Fich 2011), 'work-related communication technology use outside of regular work hours' (Wright et al. 2014) or 'daily smartphone use after working hours' (Derks et al. 2014b). Furthermore, Fenner and Renn (2010) introduced the specific term 'technology-assisted supplemental work (TASW)' to describe people who perform work tasks at home after regular work hours using technological tools such as laptops. We consider all the terms listed above as appropriate to term our research topic. However, the last listed seems to be the best for displaying the immanent characteristics of work-related use of ICT during after-hours. This choice may be clearer by considering the definition of TASW. So, according to Fenner and Renn (2010), TASW is characterized by:

– supplemental work,
– performed at home,
– after regular working hours,
– by white-collar workers,
– full-time employed,
– without a formal contract, including a compensation agreement,
– performed with new information and communication technologies, such as laptops, smartphones, and tablets.

Thus, the term TASW includes even the assumption of *supplemental* working which may be an important factor for our core outcomes employees' recovery and well-being, particularly as it refers to "any work activity during any planned free time" (Arlinghaus and Nachreiner 2014, p. 2). Thus, examples of TASW include work-related mobile communication (e.g., making and receiving calls, reading and sending e-mails, communicating via mobile instant messaging software), using mobile office functions (e.g., word processing, spreadsheets, presentation software, calendar, address book, and calculator), and work-related mobile information searches (see, e.g., Yuan et al. 2010; Yun et al. 2012). Thereby, in view of the contextual fit of this term to our purposes as well as of the usefulness of its acronym, in the following text we will use also the term TASW for reference to our research topic. However, we focus only the use of new technologies with internet access. Furthermore, the supplemental work may be performed also apart from home.

In the following section, the overall view of the operationalization of TASW follows. Table 4.1 presents specific information on the operationalization of TASW in the studies included in our review.

Table 4.1 Operationalization of TASW

No.	Study	Sample	Design	Measure of TASW
1.	Barber and Jenkins (2013)	N = 315 USA MTurk; full-time workers	Cross-sectional Online survey	Frequency of boundary crossing with ICTs: "how often one tries to arrange, schedule or perform job-related activities outside of normal work hours using ICTs" and "how often ICTs are used to perform one's job when one is at home during non-work hours" Likert scale from 1 (never/almost never) to 5 (very often/ almost always)
2.	Berkowsky (2013)	N = 865 USA Employees	Cross-sectional Web-based survey	Frequency of work-related ICT activities while at home: "how often the respondent checked work email at home, how often the respondent texted work colleagues while at home, and how often the respondent communicated with work colleagues using FB [Facebook] while at home" Response choices: from 0 (never engaged in the ICT activity) to 4 (engaged in the activity more than once a day)
3.	Boswell and Olson-Buchanan (2007)	N = 360 employees N = 35 significant-others USA Nonacademic staff employees of a public university	Cross-sectional Survey packets	Frequency of use an array of CTs (cell phones, e-mail, voice mail, PDAs, and pagers) to perform job during non-work hours Likert scale from 1 (never) to 5 (very often, i.e., several times a day)
4.	Chen and Karahanna (2014)	N = 16 and 137 USA Employees at a technology firm	Cross-sectional Interviews and web-based survey	Frequency of overall work-to-non-work interruptions (WTN): "During nonwork hours, how frequently are you interrupted by colleagues/other work contacts about work-related matters [WTN-overall] overall through technologies such as phone call, email, IM, texting etc.? Composite: composite index created by the following items: [WTN-phone] via phone call only? [WTN-email] via email only? [WTN-IM] via IM only? [WTN-texting] via texting only?"

(continued)

Table 4.1 (continued)

No.	Study	Sample	Design	Measure of TASW
5.	Chesley (2005)	N = 1.367 USA Working individuals; participants of Cornell Couples and Career study	Longitudinal Telephone survey	Frequencies of computer-based (email and internet) and communications technology (cell phones and pagers) use Regular technology use: At Time 1 dichotomous variable: regular/not regular use At Time 2 Likert scale from 1 (never use) 5 (use a lot) Persistent technology use: use at both time periods (i.e., regular use at Time 1 or a use frequency of 2–5 at Time 2)
6.	Chesley (2006)	N = 581 couples USA Working individuals; participants of Cornell Couples and Career study	Longitudinal Telephone survey	See operationalization by Chesley (2005) in the line above
7.	Derks and Bakker (2012)	N = 69 Full-time employees in possession of a smartphone for business purposes; did not use their smartphone on their own initiative (their employer initiated it)	Diary online questionnaire on five successive workdays	Extent of smartphone use: e.g., "I use my smartphone intensively", "When my smartphone blinks to indicate new messages, I cannot resist checking them" Likert scale from 1 (totally disagree) to 5 (totally agree)
8.	Derks et al. (2014a)	N = 80	Diary online questionnaire on six workdays equally spread over 2 weeks	Dichotomous variable: smartphone/non-smartphone group (PC-group)
9.	Derks et al. (2014b)	N = 100 Netherlands Full-time employees	Diary online questionnaire on four successive workdays	Extent of daily smartphone use: e.g., "Today, I used my smartphone intensively", "When my smartphone blinked to indicate new messages, I could not resist to check them today" Likert scale from 1 (totally disagree) to 5 (totally agree)

(continued)

Table 4.1 (continued)

No.	Study	Sample	Design	Measure of TASW
10.	Derks et al. (2014c)	N = 70 Germany Employees using smartphones on initiative of their employer	Diary online questionnaire on four successive workdays	Extent of daily smartphone use: e.g., "When my smartphone blinked to indicate new messages, I could not resist checking them" and "Today, I was online until I went to sleep." (see also Derks et al. 2014b) Likert scale from 1 (totally disagree) to 5 (totally agree)
11.	Diaz et al. (2012)	N = 193 Southern US Non-academic managers working at a large university	Cross-sectional Survey	Extent of ICT use for work-related purposes at home: e.g., "To what extent do you use communication technology to perform your job during non-work hours?"
12.	Fenner and Renn (2010)	N = 227 USA Full-time employees	Cross-sectional Surveys with stamped self-addressed envelopes	Frequency of TASW: e.g., "I perform job-related tasks at home at night or on weekends using my cell phone, pager, BlackBerry® or computer." Likert scale from 1 (never) to 5 (always)
13.	Glavin and Schieman (2010)	N = 1.090 USA Participants who were participating in the paid labor force	Cross-sectional Telephone interviews	Frequency of role blurring at home: "How often do coworkers, supervisors, managers, customers, or clients contact you about work-related matters outside normal work hours? Include telephone, cell phone, beeper and pager calls, as well as faxes and e-mail that you have to respond to" Response choices: 1 (never), 2 (less than once a month), 3 (once a week), 4 (several times a week), and 5 (once or more times a day)
14.	Glavin and Schieman (2012)	N = 1.075 USA	Cross-sectional Telephone interviews	See operationalization by Glavin and Schieman (2010) in the line above
15.	Glavin et al. (2011)	N = 1.042 USA Participants who were participating in the paid labor force	Cross-sectional Telephone interviews	See operationalization by Glavin and Schieman (2010) in the line above

(continued)

Table 4.1 (continued)

No.	Study	Sample	Design	Measure of TASW
16.	Lanaj et al. (2014)	N=82 Midwest US Mid- to high-level managers enrolled in weekend MBA classes at a large university	Diary online questionnaires on ten successive workdays	Duration of late-night smartphone use for work: "How many minutes did you use your Blackberry/Smartphone for work after 9 PM last night?"
17.	Lanaj et al. (2014)	N=136 MTurk	Diary online questionnaires on ten successive workdays	Duration of late-night smartphone use for work Duration of late-night desktop/laptop computer use for work Duration of late-night electronic tablet use for work See operationalization by Lanaj et al. (2014) in the line above
18.	Leung (2011)	N=612 Hong Kong Full-time office workers whose jobs require the use of the Internet	Cross-sectional Telephone survey	Dichotomous variable activity scope of ICT use for work-related purposes at home: "Besides e-mail, do you use IM, chat rooms, blogs, Web surfing, and on-line news to do office work at home?"
19.	Nam (2014)	N=850 USA Employed people who use the Internet	Cross-sectional Telephone interviews	Frequency of ICT use for work-related purposes at home: "How often do you work from home? (Based on a respondent's willingness and usage of technologies such as Internet, email, cell phones, instant messaging)" Response choices: 1 (never), 2 (less often), 3 (a few times a month), 4 (a few times a week), 5 (almost every day), and 6 (every day)
20.	Ohly and Latour (2014)	N=1.714 Germany Employed or self-employed individuals	Cross-sectional Online survey	Dichotomous variable: use of smartphones for work in the evening = SUWE, non-use = non-SUWE
21.	Olson-Buchanan and Boswell (2006)	N=938 USA Non-academic university staff employees	Cross-sectional Field survey via internal mail system	Setting boundaries for use of communication technologies to perform work during non-work time: e.g., "I limit the amount of time or when I use communication technologies for work purposes during non-work hours (for example, only until 7 p.m.)" and "I do not use communication technologies for work purposes on weekends" Dichotomous answers; items were summed to an index

(continued)

Table 4.1 (continued)

No.	Study	Sample	Design	Measure of TASW
22.	Park and Jex (2011)	N = 281 USA Full-time employees working in an office setting	Cross-sectional Online survey	Boundary creation around ICT use: "I do not use communication/information technologies for work purposes during the weekends" Likert scale
23.	Park et al. (2011)	N = 269 USA Full-time employees University alumni	Cross-sectional Online survey	Frequency of ICT use for work-related purposes at home during non-work hours: respondents were asked to report the frequency with which they use an array of communication technologies for work-related purposes at home during non-work hours Likert scale from 1 (almost never) to 5 (very often)
24.	Richardson and Benbunan-Fich (2011)	N = 139 Northeastern US Members of the marketing division of a media organization in a large metropolitan city Full-time employees	Cross-sectional Online questionnaire	Overall index of work-related connectivity behavior during non-work time (WCBA) based on frequency of ICT use: (1) asking about the use of specific technological devices (e.g., wireless enabled device), (2) asking how frequently each device is used during a specific non-work activity or event (e.g., shopping, traveling, dinner with friends, etc.)
25.	Richardson and Thompson (2012)	N = 139 Northeastern US Members of the marketing division of a media organization in a large metropolitan city Full-time employees	Cross-sectional Online questionnaire	Duration of work-related connectivity behavior during non-work time (WCBA): asking about the average time participants use specific technological devices (e.g., wireless email devices and laptops) to perform job-related duties during non-work hours (ranges of minutes such as 1–30 min, 31–60 min) Frequency of work-related connectivity behavior during non-work time (WCBA): asking how frequently each device (e.g., handheld wireless devices, laptops) is used during a specific non-work activity or event (e.g., shopping, traveling, dinner with friends, etc.) See also Richardson and Benbunan-Fich (2011)

(continued)

Table 4.1 (continued)

No.	Study	Sample	Design	Measure of TASW
26.	Schieman and Glavin (2008)	N=2.671 USA Participants who were participating in the paid labor force	Cross-sectional Telephone interviews	Frequency of role blurring at home: "How often do coworkers, supervisors, managers, customers, or clients contact you about work-related matters outside normal work hours? Include telephone, cell phone, beeper and pager calls, as well as faxes and e-mail that you have to respond to" Response choices from 1 (never) to 8 (many times a day) See also Glavin and Schieman (2010)
27.	Senarathne Tennakoon et al. (2013)	N=425 Canada Managers and professionals	Cross-sectional Web-based survey	Duration of ICT use for work-related activities: e.g., "Think of a typical WORKING day and a NONWORKING day during the last week. Give the best possible estimate for the number of hours spent using each of the following technologies: […] Using e-mail for work-related activities. […] Using Internet (other than for email access) for work-related activities. […] Portable communication devices (Cell phone/Blackberry/PDA) for work-related activities" Ranges of minutes such as 1 (none), 2 (less than 1 h), 6 (more than 5 h)
28.	Wajcman et al. (2010)	N=850 Australia Full-time and part-time employees	Cross-sectional Online/offline questionnaire and time diary	Duration of Internet use for work purposes during non-work time: combination of questionnaire and time diary data
29.	Wajcman et al. (2008)	N=1.358 individuals (877 employees)	Cross-sectional Web-based survey	Phone-logs; number of calls made and received
30.	Wright et al. (2014)	N=168 Midwest US Full-time employees	Cross-sectional Online survey	Duration of work-related communication technology use outside of regular work hours: estimating the number of minutes/hours spent each week communicating via technologies with supervisors, coworkers, or performing work-related tasks (e.g., answering questions, talking to clients, sharing files) via technology outside of regular work hours

Note: Detailed information is provided if available from the studies

As can be seen in Table 4.1, to inform better on the scope of the field, we have taken into consideration also some articles that cover a broader range of technology use (see, e.g., Arlinghaus and Nachreiner 2013; Arlinghaus and Nachreiner 2014; Day et al. 2012; Glavin and Schieman 2010; Glavin et al. 2011; Glavin and Schieman 2012; Leung 2011; Schieman and Glavin 2008; Schieman and Young 2013). Furthermore, as can be noted, we made some exceptions by the inclusion of the criterion 'non-work' reference noted above (see No. 7–8, 10–13, and 34). These articles have been included because of special research design or findings: For example, Derks and Bakker (2014) and Derks et al. (2014b, 2015) did not exactly ask about using the smartphones for work purposes. Nevertheless, they presumed it due to the inclusion criterion of using a business smartphone where the employer fully covers the costs. Furthermore, Derks et al. (2014a) distinguish only between group of smartphone users and a control group (PC-group). In this context, it should be emphasized that all these mentioned studies by Derks and her colleagues used diary design, which is a state of the art within the methods of contemporary work and organizational psychology (see Ohly et al. 2010). The studies by Chesley (2005, 2006) were included due to the longitudinal approach and the implicit assumption of TASW as the technology use was captured through their regularity as well as persistency. At last, the study by Wajcman et al. (2008) found our consideration due to three reasons: First, it investigates the overall use of mobile phones (including after hours), second, it captures ICT use more objectively than the other studies listed above, and third, it presents the interesting finding that number of calls made and received each day is not substantially related to the work-family spillover, but the preferred weekly work hours (fewer than the participants did) are.

In the following, we describe the previous operationalization of TASW. In general, work-related ICT use after hours has been operationalized in several ways. Most frequently, the participants were asked to report *how often* they use communication technologies for work-related purposes during non-work time (e.g., Boswell and Olson-Buchanan 2007; Park et al. 2011). Another common way to capture TASW is to ask participants to report their perceived *extent* of use; Diaz et al. (2012), for example, have done so: "To what extent do you use communication technology to perform your job during non-work hours?" (p. 504). Furthermore, the 4-item intensive smartphone-use scale by Derks and Bakker (2014) was used in the empirical research to measure the extent or intensity of participants' use. In this scale, smartphone use is considered as a habit/trait. Example items are "I use my smartphone intensively" and "When my smartphone blinks to indicate new messages, I cannot resist checking them" (p. 421). The scale is in our view conceptually comparable with the concept of workplace telepressure by Barber and Santuzzi (2015), in particular with the subscale urge. They defined telepressure as "the combination of a strong urge to be responsive to people at work through message-based ICTs with a preoccupation with quick response times" and captured it, for example, with the following item: "It's difficult for me to resist responding to a message right away" (p. 18). Furthermore, Derks et al. (2015; see also Derks et al. 2014b) assume daily fluctuations in the use of smartphones and, therefore, they have adjusted the scale for daily measures: "Today, I used my smartphone intensively" and "When my smartphone blinked to indicate new messages, I could not resist to check them today" (p. 9). As far as we know, the *duration* of TASW has been measured similarly often as the extent of use. In such cases, participants had to indicate time they spent on technology use. In particular, Lanaj et al. (2014) asked them: "How many minutes did you use your Blackberry/Smartphone for work after 9 PM last night?" (p. 14). Likewise, Wright et al. (2014) used this way to capture the technology use and asked for estimating time spent communicating via technologies with supervisors, coworkers, or performing work via technology outside of regularly defined work hours. A *dichotomous* measurement of TASW was used rarely. For example, Ohly and Latour (2014) captured whether participants used smartphones for work purposes and if they did so during the evening. Thereafter, the authors formed groups of users and non-users. Some researchers recorded use of *specific devices*. For example, Lanaj et al. (2014) assessed the use of computer, tablet, and television in addition to the smartphone use. More specifically, Ohly and Latour (2014) and Yun et al. (2012) asked only about the use of smartphones. Richardson and Benbunan-Fich (2011) and Richardson and Thompson (2012) captured the use of many technological devices (e.g., wireless enabled device) as well as how frequently each device was used during a specific non-work activity or event such as shopping, traveling, and dinner with friends. Actually, only a few authors were interested especially in recording of non-work *user contexts* (see, among others, Wajcman et al. 2010). Another option to conceptualize TASW is to use measures of *boundary strategies* as a proxy. For instance, Olson-Buchanan and Boswell (2006) asked about limiting the amount of time when technology is used for work purposes during non-work hours: "I limit the amount of time or when I use communication technologies for work purposes during non-work hours (for example, only until 7 p.m.)" and "I do not use communication technologies for work purposes on weekends." What these measures have in common is the fact that they are self-reported. Thus, until now, *objective measures* of technology use do not appear to be currently used for research purposes in our context. However, the rapid development and dissemination of new technology suggests future changes towards this field in the next time. For example, Wajcman et al. (2008) have already assessed accurate phone logs of communications traffic by drawing on the information already stored in handsets, combined with self-ratings, to provide a precise and comprehensive record of the telephonic activity (connections with family, friends, and colleagues).

In the context of measuring TASW, it is interesting to note that depending on the operationalization, technology use seems to be differently linked with outcomes. For example, the use of smartphones was positively related to detachment, but the duration (or intensity) was negatively linked to psychological detachment from work (Ohly and Latour 2014). Moreover, only the persistence of technology use, but not the use per se was linked to negative affect and lower family satisfaction in the longitudinal study (N =1.367) on technology use and blurring boundaries by Chesley (2005). Thereby, technology use was captured through asking for frequency of computer- and internet use (at time 1 as a dichotomous variable regular/not regular use, at time 2 with a Likert scale from 'never use' to 'use a lot'). Persistent technology use was operationalized as the use at both time periods (such as regular use at time 1 or a use frequency of 2–5 at time 2). Similarly, the specific indirect effect of work-related ICT use after hours on work interference with family through job control was statistically significant for ICT duration, but not for ICT frequency (Richardson and Thompson 2012). These results indicate that care needs to be taken when comparing results across studies to prevent false conclusions. Furthermore, it might be wise to include multiple operationalizing technology use in future studies.

4.2   Antecedents of Work-Related ICT Use After Hours

There are nine studies that consider the antecedents of TASW and meet our inclusion criteria. Table 4.2 summarizes the relevant studies.

Table 4.2 Studies examining predictors of TASW

Study	Predictors of TASW	Indicators of TASW
Chesley (2006)	Demographics, workplace and job context, family context, e.g.: Manufacturing sector [+] Work hours [+] Control over scheduling [+] Work load [+, only men] Work performance [+, only men]	Frequencies of computer-based (email and Internet) or communications technology (cell phones and pagers) use
Fenner and Renn (2010)	H1: Perceived usefulness of technology [+] H2: Organizational expectations (psychological climate for TASW; e.g., "How would you rate your employer's efforts to measure and track employees' use of technological tools to work from their homes at night or on weekends?") [+]	Frequency of TASW
Glavin and Schieman (2010)	Workplace interpersonal conflict [+] Workplace social support [+, contradictory result against assumption]	Frequency of role blurring at home
Glavin and Schieman (2012)	Schedule control [+] Decision-making latitude [+] Job authority (e.g., Do you have the authority to hire or fire others?) [+] Work pressure [+] Higher-status work aspirations [+, ns]	Frequency of role blurring at home
Olson-Buchanan and Boswell (2006)	H3: Work to non-work role permeability (e.g., "I deal with nonwork issues while at work as needed") [−] H3: Role-referencing (e.g., "I talk about my work life with my friends and family") [−]	Setting boundaries for use of communication technologies to perform work during non-work time

(continued)

Table 4.2 (continued)

Study	Predictors of TASW	Indicators of TASW
Richardson and Benbunan-Fich (2011)	H1: Organizational distribution of wireless enabled devices [+] H2: Subjective norms about after-hours work connectivity [+] H3: An organizational member's polychronicity [+] H4: An organizational member's role integration preference [+, ns] H5: An individual's level of personal innovativeness with information technology [+, ns]	Overall index of work-related connectivity behavior during non-work time (WCBA) based on frequency of ICT use
Schieman and Glavin (2008)	Men [+] Schedule control and job autonomy [+], moderator sex (among men only)	Frequency of role blurring at home
Senarathne Tennakoon et al. (2013)	H1: Perceived utility of ICT [+] H2: Segmentation [−] H3: Work flexibility [+] H4: Time pressure [+]	Duration of ICT use for work-related activities

Note: "+" denotes a positive effect, "−"denotes a negative effect, "ns" means "not significant". We present items examples if available/given in the listed studies

As can be seen in Table 4.2, the empirically investigated antecedents of supplemental work through ICT use at home vary. In brief, the antecedents are related to *technology* (e.g., perceived usefulness of technology), *work* (e.g., work flexibility), *organization* (e.g., organizational expectations), *individual* (e.g., subjective norms) and *behavior* (e.g., segmentation). This broad range of factors points to the contribution of *personal* (e.g., polychronicity) as well as *situational* (e.g., time pressure) respectively *environmental* (e.g., organizational distribution of devices) factors for the work-related ICT use after hours. Individuals who prefer to integrate their roles and to work on multiple activities at the same time use ICT more often or with higher frequencies. ICTs are used more in work contexts that allow individuals to schedule their work and are demanding. For example, Arlinghaus and Nachreiner (2014) found that employees with trust based (entirely self-determined) working hours had worked more often outside of regular working hours than those with work hours determined by their employer. Furthermore, in a study among full-time working adults in the marketing division of a media organization, Richardson and Benbunan-Fich (2011) have shown the importance of both personal and environmental factors. Thereby, they investigated the organizational and individual antecedents of work-related connectivity behavior during non-work time (WCBA). WCBA was related to the distribution of wireless enabled devices by the organization and organizational norms about connectivity. Interestingly, individual characteristics were related to WCBA depending on the functionality of the device. For example, polychronicity, as an individual characteristic, was more strongly related to laptop WCBA than to handheld WCBA, whereas role integration preference was only related to handheld WCBA but unrelated to non-handheld, such as laptop WCBA.

According to the 'locus of causality' (Deci and Ryan 2000, see Sect. 2.3), we may distinguish also between *internal* (e.g., autonomous motivation) and *external* (e.g., controlled motivation, organizational expectations) determinants of ICT use (see also). In this context, Ohly and Latour (2014) yielded similar moderate correlations for autonomous as well as for controlled motivation and ICT use.

Further predictors are listed in the section below which presents the mediating effects of ICT use (see Sect. 4.4).

4.3   Consequences of Work-Related ICT Use After Hours

We found 29 studies that consider the outcomes of TASW and meet our inclusion criteria. The outcomes of TASW vary across the studies. Table 4.3 summarizes the relevant studies.

Table 4.3 Studies examining outcomes of TASW

Study	Outcomes of TASW	Indicators of TASW
Barber and Jenkins (2013)	H1: Psychological detachment –> sleep quantity, moderator boundary creation [supported] H2: Psychological detachment –> sleep quality, moderator boundary creation [supported] H3: Psychological detachment –> sleep consistency, moderator boundary creation [supported]	Frequency of boundary crossing with ICTs
Berkowsky (2013)	H1: Negative work-home spillover [+, exception: FB seemed to prevent or lessen negative spillover]	Frequency of work-related ICT activities while at home via email, texting, Facebook
Chen and Karahanna (2014)	Work-life conflict [+] Non-work performance (fulfillment of personal life responsibilities) [−]	Frequency of overall work-to-non-work interruptions (WTN)
Chesley (2005)	H1: individual distress (negative affect) and family satisfaction through increased levels of spillover (primarily negative work-family spillover)	Persistent communications use
Derks and Bakker (2012)	H4: Work–home interference [+]	Extent of smartphone use
Derks et al. (2014a)	H1: Work-home interference [+, ns]	Dichotomous variable (smartphone/non-smartphone group)
Derks et al. (2014b)	H1: Daily work-home interference [+] H2: Daily work-home interference, moderator expectations by supervisor [supported] H3: Daily work-home interference, moderator expectations by colleagues [ns] H4: Daily work-home interference, moderator work engagement [supported]	Extent of daily smartphone use

(continued)

Table 4.3 (continued)

Study	Outcomes of TASW	Indicators of TASW
Derks et al. (2014c)	H1: Daily psychological detachment [−] H2b: Daily psychological detachment –> daily work-related exhaustion [+, partially] H3: Psychological detachment, moderator perceived workplace segmentation norm [ns] => Overall a small but significant effect providing evidence for proposed model	Extent of daily smartphone use
Fenner and Renn (2010)	H3: Work-to-family conflict [+] H4: Moderator time management (setting goals and priorities) [supported]	Frequency of TASW
Glavin and Schieman (2010)	Work-to-family conflict [+]	Frequency of role blurring at home
Glavin and Schieman (2012)	Work-to-family conflict [+] Work-to-family conflict [+], moderators schedule control [partially supported], decision-making latitude [supported], excessive work pressures [supported]	Frequency of role blurring at home
Glavin et al. (2011)	Feelings of guilt [+, ns] Distress [+], moderator sex (among women only) Feelings of guilt and distress, moderator sex [+] (strongly among women)	Frequency of role blurring at home
Lanaj et al. (2014)	H1: Sleep quantity [−] H3: Sleep quantity –> morning depletion H4: Sleep quantity and morning depletion –> daily work engagement	Duration of late-night smartphone use for work
Lanaj et al. (2014)	Use of smartphones and computers each explained more variance in sleep quantity (than tablet and TV)	Duration of late-night smartphone use for work Duration of late-night desktop/laptop computer use for work Duration of late-night electronic tablet use for work
Leung (2011)	Work spillover into home, job burnout, job satisfaction, family satisfaction [ns]	Dichotomous variable activity scope of ICT use for work-related purposes at home
Nam (2014)	H3a: Job satisfaction [ns] H3b: Job stress [−] H3c: Workload [+]	Frequency of ICT use for work-related purposes at home
Ohly and Latour (2014)	H1: Positive affect [−], detachment [+, contradictory result against assumption], recovery [−, ns], negative affect [+, ns] H2: Autonomous motivation for smartphone use –> positive affect [+], detachment [+], recovery [−], negative affect [−] H3: Controlled motivation for smartphone use –> positive affect [−], detachment [−, ns], recovery [−, ns], negative affect [+].	Dichotomous variable use of smartphones for work in the evening

(continued)

Table 4.3   (continued)

Study	Outcomes of TASW	Indicators of TASW
Richardson and Thompson (2012)	H1: Work interference with family [+] and well-being [−, ns] H2a: Job control [+] −> work interference with family [−, supported for WCBA duration but not for WCBA frequency] H3a: Detachment from work [−] −> work interference with family [−] H4: The specific indirect effect of detachment from work as a mediator between WCBA and work interference with family will be greater than the specific indirect effect of job control as a mediator [supported]	Duration of work-related connectivity behavior during non-work time (WCBA) Frequency of work-related connectivity behavior during non-work time (WCBA)
Schieman and Glavin (2008)	Work-to-home conflict [+], moderator job autonomy (only among workers with low job autonomy)	Frequency of role blurring at home
Wajcman et al. (2010)	Work-family spillover [−, contradictory result against assumption]	Duration of Internet use for work purposes during non-work time
Wajcman et al. (2008)	Work-family spillover [ns] Family-work spillover [+]	Number of calls made and received
Wright et al. (2014)	H1: Work-life conflict associated with communication technology use [+]	Duration of work-related communication technology use outside of regular work hours

Note: "+" denotes a positive effect, "−"denotes a negative effect, "ns" means "not significant"

   In short, the examined outcomes may be roughly classified as covering recovery indicators (e.g., sleep quantity and psychological detachment) as well as subjective well-being (e.g., positive and negative affect, job satisfaction, job burnout).

   Recovery has already been identified as a core concept in the research on ICT use. For example, Barber and Jenkins (2013) revealed negative effects on *sleep* for individuals with low boundaries around technology use. In recent research, the use of light-emitting electronic devices immediately before bedtime is examined. In this context, Lanaj et al. (2014) investigate the consequences of smartphone use for work at night. Two diary studies showed that smartphone use for work at night reduced sleep quantity, which in turn increased morning depletion and this process had negative consequences for daily work engagement. Concerning the adverse consequences for sleep, the authors argue with changes in physiological (reduced melatonin secretion) as well as psychological processes (reduced psychological detachment from work) through smartphone use. The effect of work-related ICT use during after work hours on *psychological detachment* has already been investigated several times (Barber and Jenkins 2013; Derks et al. 2014b; Ohly and Latour 2014). In contrast, the *physiological effects* of ICT use have been assumed only by Lanaj et al. (2014). Hence, the research on them has increased recently (e.g., Chang et al. 2014; Sroykham and Wongsawat 2013; Wood et al. 2013). Moreover, there are already empirical results demonstrating that evening exposure to light-emitting electronic devices extends the time it takes to fall asleep, disrupts the circadian rhythm, acutely suppresses sleep hormone melatonin, modifies REM sleep, and reduces morning alertness (Chang et al. 2014). Interestingly, in the study by Lanaj et al. (2014), the detrimental effects of late-night smartphone use on sleep, depletion, and engagement were incremental to the effects of other electronic devices such computer, tablet, and television. Furthermore, the detrimental effects of computer use for work late at night on the outcomes were very robust. The arguments for those phenomena were related to the small size and proximity of smartphones– their impact on sleep may be primarily psychological (through lack of psychological detachment), whereas the larger displays of computer may be primarily biological (through melatonin suppression).

In the context of *well-being*, Lanaj et al. (2014) have already associated recovery with well-being indicators as they examined also the effects on daily work engagement. This relationship was moderated by job control. This means that morning depletion diminishes daily work engagement only for employees who perceive low job control. Taken together, in regard to well-being classification by Diener et al. (1999, see Sect. 3.2.1), not all four components of well-being have been investigated. For example, to our knowledge, *life satisfaction* has not yet been investigated in the context of work-related ICT use after hours. In regard to *domain satisfactions*, most studies have examined the relationship between ICT use and indicators of boundary management, such as work-home interference, work-home-spillover or the assumed negative outcome of ICT use-work-home/life-conflict (e.g., Derks and Bakker 2014; see also the effects of computer-supported supplemental work-at-home by Duxbury et al. 1996). Job satisfaction was seldom investigated (see Nam 2014; Wright et al. 2014). As referred above, work engagement was examined by Lanaj et al. (2014). Furthermore, Yun et al. (2012) and Nam (2014) discussed the role of ICT use for the overall workload. This topic was often overlooked in the previous research, but research points to the fact that ICT might be a symbol rather than a cause of workload (Barley et al. 2011). The health component of well-being has been considered in the studies of Richardson and Thompson (2012), Arlinghaus and Nachreiner (2013), and Arlinghaus and Nachreiner (2014). In regard to *affective* consequences of ICT use, Ohly and Latour (2014) have addressed the relationships between ICT use and affective states as potential indicators of well-being. Work-related smartphone usage in the evening was positively related to psychological detachment, but unrelated to recovery and negative affect, and negatively related to positive affect in a large sample of working individuals (N = 1.706). The study by Glavin et al. (2011) on a large American population (N = 1.042) showed that the frequency of receiving work-related contact outside of normal work hours was positively associated with feelings of guilt, but among women only.

In brief, both of our core concepts–recovery as well as well-being–were already partially investigated in the context of work-related ICT use after hours. All in all, the negative associations between ICT use and recovery and well-being indicators are mostly consistent. The necessary *conditions* for some effects (in terms of moderators) are *internal,* such as a certain sort of motivation, work engagement, and boundary creation or *external,* such as the expectations by a supervisor to stay online. These and other possible circumstances will be discussed in detail when developing our conceptual model. Thus, previous research offers an initial empirical basis for our theoretical framework. Further outcomes are listed in the following sections, which present the mediating and moderating effects of ICT use.

4.4   Work-Related ICT Use After Hours as Mediator

During our literature search, we have noted that research has started exploring the possible mediating role of TASW. We identified four studies that examine mediating effects of ICT use (TASW as mediator) and meet our inclusion criteria. Table 4.4 summarizes the relevant studies.

Table 4.4 Studies examining mediating effects of ICT use

Study	Predictors of TASW	TASW as mediator	Outcomes of TASW
Boswell and Olson-Buchanan (2007)	H1: Affective commitment [+, ns], job involvement [+], ambition [+]	Post hoc analyses: Frequency of ICT use [not supported as a mediator]	H2a: Work-to-life conflict [+] H2b: Work-to-life conflict as reported by significant other [+]
Diaz et al. (2012)	H1: Perceived communication technology flexibility [+]	H4a & H4b: Extent of ICT use for work-related purposes at home [partially supported as a mediator]	H2a: Work satisfaction [+] H2b: Work-to-life conflict [+]
Park and Jex (2011)	H1: Preference for segmenting work from the family domain [+] H2: Work-role identification [-]	H4a & H4b: Boundary creation around ICT use [partially supported as a mediator]	H3: Work-to-family interference [−]
Park et al. (2011)	Work-home segmentation preference Perceived segmentation norm	H4 & H5: Frequency of ICT use Frequency of ICT use for work-related purposes at home during non-work hours [supported as a mediator]	H3: Psychological detachment from work during non-work time [-]

Note: "+" denotes a positive effect, "−" denotes a negative effect, "ns" means "not significant"

In sum, most of the assumed mediations were at least partially supported. Park and Jex (2011) provide an insight into the process of work-life conflict prevention. In their study, the preference of boundary management led to higher boundary creation around ICT use and that, in turn, decreased employees' work-to-family interference. The study by Boswell and Olson-Buchanan (2007) is methodologically challenging as they surveyed also the significant others of participants, reducing same-source bias. Furthermore, this study presents other relevant antecedents of TASW: job involvement and professional ambition. Importantly, in the context of recovery, lack of psychological detachment due to TASW seems to be determined by an individual's boundary management preferences and perceived boundary management norms (Park et al. 2011). Moreover, it is not very common in the literature to assume positive effects of TASW. More common is the proposition of negative effects on individuals' recovery and well-being (see Table 4.3). Therefore, a further real added value to the empirical results mentioned above is the assumption of a positive relationship between TASW and work satisfaction (Diaz et al. 2012). Thus, Diaz et al. (2012) posited and examined technology use as a "double-edge sword" (p. 502) by simultaneously predicting increase in work satisfaction as well as work-life conflict.

Thus, TASW mediated the relationships between *attitudes toward technology* (e.g., perceived communication technology flexibility), *boundary management/ conditions* (e.g., segmentation preference or norm), and outcomes such as indicators of *work-life balance*, *recovery* (e.g., psychological detachment), or *attitudes toward work* (e.g., work satisfaction). These results will be considered in the final theoretical framework of recovery and well-being.

4.5 Work-Related ICT Use After Hours as Moderator

There are only two studies that examine moderating effects of ICT use (TASW as moderator) and meet our inclusion criteria. Table 4.5 summarizes the relevant studies.

As can be seen in Table 4.5, smartphone users benefited from recovery experiences such as psychological detachment and relaxation when facing work-home interference. However, benefits of intense users are higher than those of non-intense users (Derks and Bakker 2014). Furthermore, intense users experience more daily exhaustion when faced with high levels of work-home interference (Derks and Bakker 2014). Moreover, smartphone users succeed less in adopting coping recovery strategies when faced with high levels of work-home interference in comparison to a control of nonusers respectively users of PC (Derks et al. 2014a). All these results can be considered in the final theoretical framework.

Table 4.5 Studies examining outcomes of TASW

Study	Predictors	Outcomes	TASW as moderator
Derks and Bakker (2012)	H5: Daily recovery (psychological detachment and relaxation) H6: Daily work–home interference	H5: Daily work–home interference [−] H6: daily burnout symptoms (daily exhaustion [+] and daily cynicism [+])	Extent of smartphone use Moderation H5: Daily recovery is more strongly negatively related to daily work – home interference for employees referring intensively smartphone use in comparison to refrained users [supported] Moderation H6: Daily work – home interference is more strongly wpositively related to daily burnout symptoms for employees referring intensively smartphone use in comparison to refrained users [not supported for cynicism]
Derks et al. (2014a)	H2: Work-home interference	H2: Recovery strategies (psychological detachment [+], relaxation [+], mastery [+], autonomy [+])	Dichotomous variable (smartphone/ non-smartphone group) Moderation H2: The positive relationship between work-home interference and recovery strategies is weaker or even negative for smartphone users in comparison to non-users [supported]

Note: "+" denotes a positive effect, "−"denotes a negative effect, "ns" means "not significant"

4.6 Additional Assumptions

As explained above, to inform better on the scope of the field, and given the current scarcity of empirical, quantitative studies on the topic, qualitative as well as theoretical approach were additionally taken in the present review. Therefore, in this section, we present a view of four further works on TASW. Thereby, the most important including criterion was a value-added content to the above-mentioned antecedents and consequences of TASW. Table 4.6 summarizes these studies. For the mentioned reasons the list does not claim to be exhaustive.

In brief, the qualitative research has offered us further views of the possible antecedents and consequences of TASW. For example, Fenner and Renn (2004) posited a substantial role of trait conscientiousness by the explanation of TASW. Due to many well-known significant effects of conscientiousness in work and organizational psychology (see among others, Barrick and Mount 1991), it would not be surprising if this trait explained a substantial part of variance in TASW or its consequences among many of the predictors examined in the previous empirical studies. Furthermore, MacCormick et al. (2012) have delivered a new view on TASW as they found out that its various levels have various consequences for work engage-

Table 4.6 Additional research focusing on TASW

Study	Sample, design	Predictors of TASW	Outcomes of TASW
Currie and Eveline (2011)	3-stage process: 1. an anonymous on-line questionnaire (N=44), 2. interviews with a smaller sample of questionnaire respondents (N=12), 3. timed diary entries using volunteers from among those interviewed (N=9) Australia academics with young children		Work-life-balance [+/−] (benefit to the work but at a cost to family life)
Fenner and Renn (2004)	USA Theoretical framework	Personal innovativeness with information technology [+] Job involvement [+] Conscientiousness [+] Career commitment [+]	Job performance [+] Career success [+] Work-to-family conflict [+] Moderators media richness, time and boundary management
MacCormick et al. (2012)	N=21 Australia Semi-structured interviews; follow up 5 years later Senior management and two focus groups across a wide range of functions at two global investment banks	Three types of smartphone users: Hypo-connectors Dynamic connectors Hyper-connectors non-stop work	Various work engagement behaviors: Disengagement Functional engagement Disengagement
Matusik and Mickel (2011)	N=54 USA Employees Grounded theory	Reasons for adoption: Technical Mimetic Coercive	Three types of user reactions: Enthusiastic Balanced Trade-offs

Note: "+" denotes a positive effect, "−" denotes a negative effect

ment. Their findings suggest curvilinear relationship between TASW and work engagement. These results correspond with the person-environment fit approach (e.g., Edwards et al. 1998) which proposes an optimum of work demands that are facilitating well-being. It is questionable if this quadratic relationship applies to all occupational groups or is specific for the investigated sample of investment bankers. Further moderator effects of potential quadratic relationships between TASW and recovery respectively well-being outcomes will be discussed in Chap. 5. Matusik and Mickel (2011) showed in their theoretical work that users of converged mobile devices (e.g., BlackBerrys, Treos, and iPhones) experience pressure to be accessible and responsive, whereby the sources of these expectations are internal as well as external to organizations (e.g., family, friends, and society in general). They found three different reasons for the adoption of this connectivity technology (technical, mimetic, and coercive) and three various types of user reactions (enthusiastic, balanced, and trade-offs). Highly interesting are the results by Currie and Eveline (2011) that undermine the 'double-edged' effect of ICT use. Thus, employees using ICT may perceive benefits as well as disadvantages for their work-life balance. The work-part improves while the family life-part deteriorates. In sum, we gained from the qualitative research additional variables that may be considered in the theoretical framework below.

4.7   Summary of the Research Findings and Conclusion

This review presents an overview of current research on work-related ICT use during non-work time published until 2014. Our sample includes primarily quantitative studies. Additionally, we have presented some non-empirical studies on ICT use that broaden the amount of possible antecedents and consequences of TASW.

In sum, we observed that most of the research on TASW has been undertaken in North America. All studies were conducted in highly economically developed countries. While many studies were cross-sectional in nature (e.g., Barber and Jenkins 2013) and therefore conclusions about causal effects are premature, some of them used diary design (e.g., Derks et al. 2014a) which provides means to examine short-term processes and daily experiences of individuals (Ohly et al. 2010). Lanaj et al. (2014) even used experience sampling methodology and collected data at different points in time during the day. Furthermore, almost all the previous studies used self-report data (except Boswell and Olson-Buchanan 2007) and might be affected therefore by same source bias (Podsakoff et al. 2003).

As can be seen from the overview above, the research focused on investigating the consequences of TASW has quantitatively greatly increased recently (see Table 4.3). In contrast, the antecedents remain underexplored (see Table 4.2), although, in our opinion, the previous studies have discussed potentially worthwhile ideas (like attitudes towards technology, see Boswell and Olson-Buchanan 2007; between-person differences as levels of energy, see Derks et al. 2014b; or the not adequately met obligations of the paid work role, such as unfinished tasks, see Glavin et al. 2011). The examined relationships in the 37 studies are summarized briefly in Tables 4.2, 4.3, 4.4, and 4.5. As can be seen in the tables, we have found all types of associations (positive, negative, or none) of antecedents as well as consequences of TASW. We have found antecedents related to *technology* (e.g., perceived usefulness of technology), *work* (e.g., time pressure and work flexibility), *organization* (e.g., organizational expectations), and *individual* (e.g., subjective norms). Furthermore, we have roughly classified the consequences as covering *recovery* (e.g., sleep quantity and psychological detachment) as well as *subjective well-being* (e.g., positive and negative affect, job satisfaction, and job burnout). In regard to the concept of well-being by Diener et al. (1999, see Sect. 3.2.1), the investigation of life satisfaction in ICT context is still missing in previous empirical research. However, the other three components of well-being (domain satisfactions, negative affect, and positive affect) were already identified as examined outcomes of TASW. Overall, TASW has been mostly associated with adverse consequences for recovery and well-being. These negative associations have been reported mostly consistently in the reviewed studies. Furthermore, some moderators in the relationships were also found. Moreover, we could identify the first examined mediator variables in the relationship between predictors and consequences of TASW. Particularly, we examined the previous conceptualization of ICT use and concluded that it might be wise to include multiple operationalizing technology use in future studies, including objective data on use taken from logs. The mentioned research gaps will be considered when developing our conceptual framework below and in the discussion towards future research.

Regarding limitations of this chapter—as often noted in similar reviews—a selection bias could have occurred. We had a very narrow research focus and, therefore, articles that did not meet all our requirements were not included in further considerations. Secondly, a publication bias is possible as we considered only published articles in scientific journals. There might be more studies that show no significance and have not been published yet.

References

Arlinghaus, A., & Nachreiner, F. (2013). When work calls-associations between being contacted outside of regular working hours for work-related matters and health. *Chronobiology International, 30*(9), 1197–1202. doi:10.3109/07420528.2013.800089.

Arlinghaus, A., & Nachreiner, F. (2014). Health effects of supplemental work from home in the European Union. *Chronobiology International, 31*(10), 1–8. doi:10.3109/07420528.2014.957 297.

Barber, L. K., & Jenkins, J. S. (2013). Creating technological boundaries to protect bedtime: Examining work-home boundary management, psychological detachment and sleep. *Stress and Health*. doi:10.1002/smi.2536.

Barber, L. K., & Santuzzi, A. M. (2015). Please respond: Workplace telepressure and employee recovery. *Journal of Occupational Health Psychology, 20*(2), 172–189. doi:10.1037/a0038278.

Barley, S. R., Meyerson, D. E., & Grodal, S. (2011). E-mail as a source and symbol of stress. *Organization Science, 22*(4), 887–906. doi:10.1287/orsc.1100.0573.

Barrick, M. R., & Mount, M. K. (1991). The big five personality dimensions and job performance: A meta-analysis. *Personnel Psychology, 44*(1), 1–26. doi:10.1111/j.1744-6570.1991.tb00688.x.

Berkowsky, R. W. (2013). When you just cannot get away. *Information, Communication & Society, 16*(4), 519–541. doi:10.1080/1369118X.2013.772650.

Boswell, W. R., & Olson-Buchanan, J. B. (2007). The use of communication technologies after hours: The role of work attitudes and work-life conflict. *Journal of Management, 33*(4), 592–610. doi:10.1177/0149206307302552.

Chang, A.-M., Aeschbach, D., Duffy, J. F., & Czeisler, C. A. (2014). Evening use of light-emitting eReaders negatively affects sleep, circadian timing, and next-morning alertness. *Proceedings of the National Academy of Sciences of the United States of America.* doi:10.1073/pnas.1418490112.

Chen, A., & Karahanna, E. (2014). Boundaryless technology: Understanding the effects of technology-mediated interruptions across the boundaries between work and personal life. *AIS Transactions on Human-Computer Interaction, 6*(2), 16–36.

Chesley, N. (2005). Blurring boundaries? Linking technology use, spillover, individual distress, and family satisfaction. *Journal of Marriage and Family, 67*(5), 1237–1248. doi:10.1111/j.1741-3737.2005.00213.x.

Chesley, N. (2006). Families in a high-tech age: Technology usage patterns, work and family correlates, and gender. *Journal of Family Issues, 27*(5), 587–608. doi:10.1177/0192513X05285187.

Currie, J., & Eveline, J. (2011). E-technology and work/life balance for academics with young children. *Higher Education, 62*(4), 533–550. doi:10.1007/s10734-010-9404-9.

Day, A., Paquet, S., Scott, N., & Hambley, L. (2012). Perceived information and communication technology (ICT) demands on employee outcomes: The moderating effect of organizational ICT support. *Journal of Occupational Health Psychology, 17*(4), 473–491. doi:10.1037/a0029837.

Deci, E. L., & Ryan, R. M. (2000). The "what" and "why" of goal pursuits: human needs and the self-determination of behavior. *Psychological Inquiry, 11*(4), 227–268. doi:10.1207/S15327965PLI1104_01.

Derks, D., & Bakker, A. B. (2014). Smartphone use, work-home interference, and burnout: A diary study on the role of recovery. *Applied Psychology, 63*(3), 411–440. doi:10.1111/j.1464-0597.2012.00530.x.

Derks, D., ten Brummelhuis, L. L., Zecic, D., & Bakker, A. B. (2014a). Switching on and off …: Does smartphone use obstruct the possibility to engage in recovery activities? *European Journal of Work and Organizational Psychology, 23*(1), 80–90. doi:10.1080/1359432X.2012.711013.

Derks, D., van Mierlo, H., & Schmitz, E. B. (2014b). A diary study on work-related smartphone use, psychological detachment and exhaustion: Examining the role of the perceived segmentation norm. *Journal of Occupational Health Psychology, 19*(1), 74–84. doi:10.1037/a0035076.

Derks, D., van Duin, D., Tims, M., & Bakker, A. B. (2015). Smartphone use and work-home interference: The moderating role of social norms and employee work engagement. *Journal of Occupational and Organizational Psychology, 88*(1), 155–177. doi:10.1111/joop.12083.

Diaz, I., Chiaburu, D. S., Zimmerman, R. D., & Boswell, W. R. (2012). Communication technology: Pros and cons of constant connection to work. *Journal of Vocational Behavior, 80*(2), 500–508. doi:10.1016/j.jvb.2011.08.007.

Diener, E., Suh, E. M., Lucas, R. E., & Smith, H. L. (1999). Subjective well-being: Three decades of progress. *Psychological Bulletin, 125*(2), 276–302. doi:10.1037//0033-2909.125.2.276.

Duxbury, L. E., Higgins, C. A., & Thomas, D. (1996). Work and family environments and the adoption of computer-supported supplemental work-at-home. *Journal of Vocational Behavior, 49*(1), 1–23. doi:10.1006/jvbe.1996.0030.

Edwards, J. R., Caplan, R. D., & Harrison, R. V. (1998). Person-environment fit theory: Conceptual foundations, empirical evidence, and directions for future research. In C. L. Cooper (Ed.), *Theories of organizational stress* (pp. 28–67). Oxford/New York: Oxford University Press.

Fenner, G. H., & Renn, R. W. (2004). Technology-assisted supplemental work: Construct definition and a research framework. *Human Resource Management, 43*(2–3), 179–200. doi:10.1002/hrm.20014.

Fenner, G. H., & Renn, R. W. (2010). Technology-assisted supplemental work and work-to-family conflict: The role of instrumentality beliefs, organizational expectations and time management. *Human Relations, 63*(1), 63–82. doi:10.1177/0018726709351064.

Glavin, P., & Schieman, S. (2010). Interpersonal context at work and the frequency appraisal and consequences of boundary spanning demands. *Sociological Quarterly, 51*(2), 205–225. doi:10.1111/j.1533-8525.2010.01169.x.

Glavin, P., & Schieman, S. (2012). Work-family role blurring and work-family conflict: The moderating influence of job resources and job demands. *Work and Occupations, 39*(1), 71–98. doi:10.1177/0730888411406295.

Glavin, P., Schieman, S., & Reid, S. (2011). Boundary-spanning work demands and their consequences for guilt and psychological distress. *Journal of Health and Social Behavior, 52*(1), 43–57. doi:10.1177/0022146510395023.

Killion, J. B., Johnston, J. N., Gresham, J., Gipson, M., Vealé, B. L., et al. (2014). Smart device use and burnout among health science educators. *Radiologic Technology, 86*(2), 144–154.

Lanaj, K., Johnson, R. E., & Barnes, C. M. (2014). Beginning the workday yet already depleted? Consequences of late-night smartphone use and sleep. *Organizational Behavior and Human Decision Processes, 124*(1), 11–23. doi:10.1016/j.obhdp.2014.01.001.

Leung, L. (2011). Effects of ICT connectedness, permeability, flexibility, and negative spillovers on burnout and job and family satisfaction. *An Interdisciplinary Journal on Humans in ICT Environments, 7*(3), 250–267.

MacCormick, J. S., Dery, K., & Kolb, D. G. (2012). Engaged or just connected? Smartphones and employee engagement. *Organizational Dynamics, 41*(3), 194–201. doi:10.1016/j.orgdyn.2012.03.007.

Matusik, S. F., & Mickel, A. E. (2011). Embracing or embattled by converged mobile devices? Users' experiences with a contemporary connectivity technology. *Human Relations, 64*(8), 1001–1030. doi:10.1177/0018726711405552.

Nam, T. (2014). Technology use and work-life balance. *Applied Research in Quality of Life, 9*(4), 1017–1040. doi:10.1007/s11482-013-9283-1.

Ohly, S., & Latour, A. (2014). Work-related smartphone use and well-being in the evening: The role of autonomous and controlled motivation. *Journal of Personnel Psychology, 13*(4), 174–183. doi:10.1027/1866-5888/a000114.

Ohly, S., Sonnentag, S., Niessen, C., & Zapf, D. (2010). Diary studies in organizational research. *Journal of Personnel Psychology, 9*(2), 79–93. doi:10.1027/1866-5888/a000009.

Olson-Buchanan, J. B., & Boswell, W. R. (2006). Blurring boundaries: Correlates of integration and segmentation between work and nonwork. *Journal of Vocational Behavior, 68*(3), 432–445. doi:10.1016/j.jvb.2005.10.006.

Park, Y., & Jex, S. M. (2011). Work-home boundary management using communication and information technology. *International Journal of Stress Management, 18*(2), 133–152. doi:10.1037/a0022759.

Park, Y., Fritz, C., & Jex, S. M. (2011). Relationships between work-home segmentation and psychological detachment from work: The role of communication technology use at home. *Journal of Occupational Health Psychology, 16*(4), 457–467. doi:10.1037/a0023594.

Podsakoff, P. M., MacKenzie, S. B., Lee, J.-Y., & Podsakoff, N. P. (2003). Common method biases in behavioral research: A critical review of the literature and recommended remedies. *Journal of Applied Psychology, 88*(5), 879–903. doi:10.1037/0021-9010.88.5.879.

Richardson, K., & Benbunan-Fich, R. (2011). Examining the antecedents of work connectivity behavior during non-work time. *Information and Organization, 21*(3), 142–160. doi:10.1016/j.infoandorg.2011.06.002.

Richardson, K. M., & Thompson, C. A. (2012). High tech tethers and work-family conflict: A conservation of resources approach. *Engineering Management Research, 1*(1), 29. doi:10.5539/emr.v1n1p29.

Schieman, S., & Glavin, P. (2008). Trouble at the border?: Gender, flexibility at work, and the work-home interface. *Social Problems, 55*(4), 590–611. doi:10.1525/sp.2008.55.4.590.

Schieman, S., & Young, M. C. (2013). Are communications about work outside regular working hours associated with work-to-family conflict, psychological distress and sleep problems? *Work & Stress, 27*(3), 244–261. doi:10.1080/02678373.2013.817090.

Senarathne Tennakoon, K. U., da Silveira, G. J., & Taras, D. G. (2013). Drivers of context-specific ICT use across work and nonwork domains: A boundary theory perspective. *Information and Organization, 23*(2), 107–128. doi:10.1016/j.infoandorg.2013.03.002.

Sroykham, W., & Wongsawat, Y. (2013, March 3). *Effects of LED-backlit computer screen and emotional self-regulation on human melatonin production.* 35th annual international conference of the IEEE EMBS, Osaka, Japan.

Wajcman, J., Bittman, M., & Brown, J. E. (2008). Families without borders: Mobile phones, connectedness and work-home divisions. *Sociology, 42*(4), 635–652. doi:10.1177/0038038508091620.

Wajcman, J., Rose, E., Brown, J. E., & Bittman, M. (2010). Enacting virtual connections between work and home. *Journal of Sociology, 46*(3), 257–275. doi:10.1177/1440783310365583.

Ward, S., & Steptoe-Warren, G. (2013). A conservation of resources approach to BlackBerry use, work-family conflict and well-being: Job control and psychological detachment from work as potential mediators. *Engineering Management Research, 3*(1), 8. doi:10.5539/emr.v3n1p8.

Wood, B., Rea, M. S., Plitnick, B., & Figueiro, M. G. (2013). Light level and duration of exposure determine the impact of self-luminous tablets on melatonin suppression. *Applied Ergonomics, 44*(2), 237–240. doi:10.1016/j.apergo.2012.07.008.

Wright, K. B., Abendschein, B., Wombacher, K., O'Connor, M., Hoffman, M., Dempsey, M., et al. (2014). Work-related communication technology use outside of regular work hours and work life conflict: The influence of communication technologies on perceived work life conflict, burnout, job satisfaction, and turnover intentions. *Management Communication Quarterly, 28*(4), 507–530. doi:10.1177/0893318914533332.

Yuan, Y., Archer, N., Connelly, C. E., & Zheng, W. (2010). Identifying the ideal fit between mobile work and mobile work support. *Information & Management, 47*(3), 125–137. doi:10.1016/j.im.2009.12.004.

Yun, H., Kettinger, W. J., & Lee, C. C. (2012). A new open door: The smartphone's impact on work-to-life conflict, stress, and resistance. *International Journal of Electronic Commerce, 16*(4), 121–152. doi:10.2753/JEC1086-4415160405.

Chapter 5
Conceptual Framework with the Focus on Recovery and Well-Being Processes

In this chapter, we discuss the previous theories and empirical results in regard to antecedents and consequences of TASW (technology-assisted supplemental work, Fenner and Renn 2010) with the focus on recovery and well-being processes. Thereafter, we propose a conceptual overall framework of antecedents and consequences of TASW. This is presented in Fig. 5.5.

As we noted above, in our opinion, the previous consideration of TASW as 'double-edged' (demand/resource) is correct in itself but not sufficient to explain all its possible consequences for recovery and well-being processes. Thus, we see the differentiation of the three work characteristics stressors, demands, and resources, based on the action theory (Hacker 1998, 2003; Frese and Zapf 1994), as fundamental for our theoretical approach (cf. JD-R model; Bakker and Demerouti 2007; Demerouti et al. 2001). This three-way division enables to propose various associations (e.g., linear negative, linear positive, buffering, and curvilinear effects) to outcomes such as strain, recovery, and well-being (Zapf and Semmer 2004). In the context of TASW, we propose that it may serve as stressor, demand, and/or resource in the daily process of employees' recovery and well-being. The underlying circumstances of these three assumptions are discussed in the following sections. This focus on action theory does not imply that we neglect the other presented theories in the development of our theoretical model. Instead, we interconnect all of them while explaining and evaluating TASW behavior. For example, cognitive appraisals (see Lazarus and Folkman 1984) are suggested to play a central role when considering TASW as stressor, demand, or resource.

At this point, it should be noted that due to the scarcity of studies in the focused research field, unconfirmed relationships proposed by researchers do not mean that those relationships really do not exist. Therefore, these associations should not be automatically neglected in the current state of research. On the contrary, they are to be reexamined in future well-designed studies based on representative samples. Such designs will be discussed in Chap. 6. Thus, in following we present assumptions of antecedents and consequences of TASW with the focus on recovery and well-being processes based on previous theories and empirical results presented above (see Chaps. 2, 3, and 4).

© The Author(s) 2016
L. Ďuranová, S. Ohly, *Persistent Work-Related Technology Use, Recovery and Well-being Processes*, SpringerBriefs in Psychology,
DOI 10.1007/978-3-319-24759-5_5

5.1   Antecedents of Work-Related ICT Use After Hours

In this section, we present the assumptions from the theories presented above as well as from the previous research with the aim of compiling a broad range of TASW determinants for our research model. As already mentioned, a number of individual attributes (personal factors) and external factors (environmental factors) may lead to TASW. Due to the aim of this work and its focus on processes of recovery and well-being, we consider also daily job characteristics and daily states in the overall conceptual model. Figure 5.1 summarizes the proposed relationships.

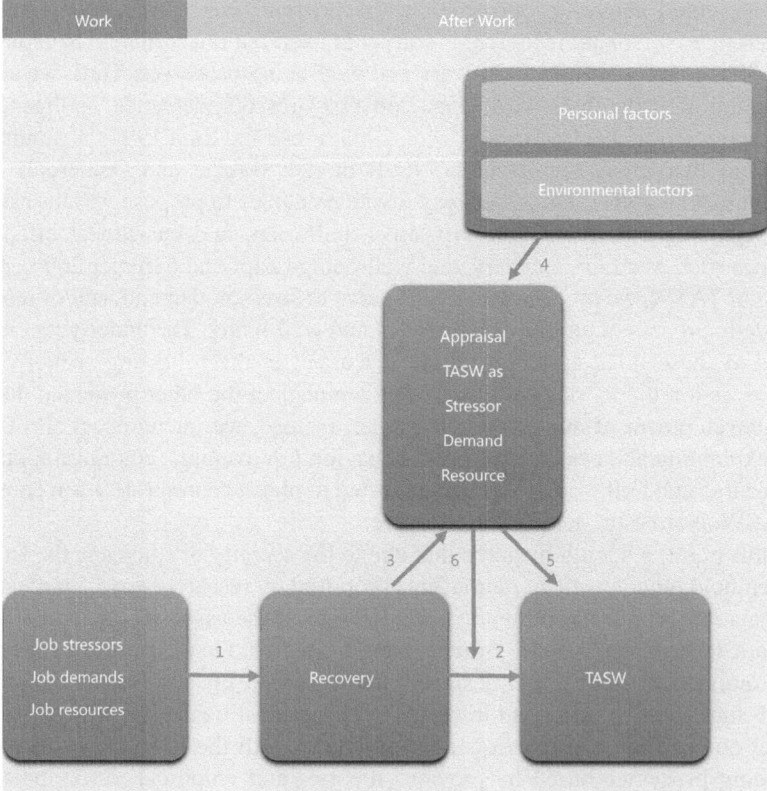

Fig. 5.1 The conceptual model of antecedents of TASW with focus on recovery and well-being processes

5.1.1 States

5.1.1.1 After Work Recovery

Due to the focus on processes of recovery and well-being, in the following we start with the daily state of recovery after work. According to the effort-recovery model (Meijman and Mulder 1998, see Sect. 3.1.1), during work employees spend effort to provide work performance. The effort expenditure causes acute physiological, behavioral, and subjective load reactions like elevated blood pressure and fatigue. As a consequence, employees need sufficient recovery from work-related effort during after-hours, otherwise the acute load reactions may have serious implications for their health and well-being. Appropriate non-work activities with recovering character do not draw on the same psychophysiological activation or use the same resources as work does (Meijman and Mulder 1998). This is in accordance with COR theory (Hobfoll 1989, 2011, see Sects. 2.4 and 3.1.1) that supposes straining resources losses and/or beneficial resources gains during work which, in turn, may have consequences for individual behavior in regard to conservation of resources. Thus, employees who avoid using ICT for work purposes during after-hours (segmentors, see Sect. 2.1), may do it in order to conserve their resources after demanding workdays. They need to refill their energy reserves by detaching and restoring from the job demands. According to the self-control model (Baumeister et al. 1998; Muraven and Baumeister 2000, see Sect. 3.1.1), employees experience draining of their self-control resources during work. These resources are vulnerable to depletion and depletion can be reversed by the replenishment of resources (e.g., through glucose intake or by rest and relaxation). In particular, knowledge workers who may use high levels on executive functions at work may be highly affected by ego depletion and therefore have high need for recovery from work-related tasks during leisure time.

In sum, according to the presented theories, we expect that characteristics of a work day (perceived *as job demands*, *job stressors*, or *job resources*; see, e.g., Sect. 2.8 and Hackman and Oldham 1980) may be substantial predictors of the individual level on recovery indicators (such as need for recovery or capacity of self-regulation) after work [Path 1, Fig. 5.1]. The influence of daily job characteristics on employee recovery level after hours will be not described in detail as it presents one of the basic assumptions in work psychology. In this section, we concentrate on the core concept of TASW. Thus, our further assumption is that the after work recovery may determine if and how much employees will work during leisure time through ICT [Path 2, Fig. 5.1]. If employees are depleted, they will avoid TASW and try to recover from daily work demands and stressors. Furthermore, this state may influence their appraisals of potential TASW [Path 3, Fig. 5.1] as described in the section below.

5.1.1.2 Appraisals of Work-Related ICT Use After Hours

At this point, we present ideas why some employees may want to continue their work through ICT after hours. As noted above, employees need to restore their energy resources after work in dependence on their current level of energy.

Therefore, at first (before showing TASW behavior), during the *primary appraisal process* (Lazarus and Folkman 1984; see details in Sect. 2.5), they should appraise the potential continuous work through ICT after hours as neutral, beneficial, or stressful. This process includes an evaluation regarding effects of TASW on one's recovery and well-being.

First, appraising TASW as beneficial is in our view conceptually equal to the *resource* concept. In this case, employees see TASW as helping to attain their goal (e.g., in employees high on professional ambition). We suppose that employees who appraise TASW as beneficial or as a resource would more likely show this behavior during leisure time. Second, appraising TASW as causing loss or harm after experiences with it is in our view conceptually equal to the concept of *stressor*. We expect that employees who have already experienced TASW as a stressor which drained their energy resources are more likely to avoid showing this behavior in the future. The third possible appraisal of TASW is based on the expectation that it will be a threat or a challenge for employees. Both of them can, in our view, be called *demand*. The most important differentiation between stressor and demand is that demand is not necessarily related to negative outcomes (see Bakker and Demerouti 2007; Zapf and Semmer 2004; for more detailed argumentation see Sect. 5.3). Furthermore, combining the transactional model of stress (Lazarus and Folkman 1984) with the challenge-hindrance framework (Cavanaugh et al. 2000; LePine et al. 2005, see Sect. 2.7), we can distinguish between 'challenge demands' and 'hindrance demands', as the concept of stressor will be reserved only for very specific contexts (see Sect. 5.3.1). For example, TASW may be appraised as a *challenge demand* (like time pressure) if its consequences will be associated with overcoming of certain obstacles (the 'straining part') to attain a valuable goal (the 'resourcing part'). On the other hand, TASW may be appraised as a *hindrance demand* if its consequences are only negative connoted (only the straining part). The detailed distinction between challenge demands and hindrance demands will be presented in Sect. 5.3.3 on TASW as a demand below.

During the *secondary appraising*, employees judge their coping resources (e.g., skills or social support) that might help in dealing with the stressful situation. After the both appraisal processes follows *re-appraisal* which means a new appraisal, a re-evaluation of the event. For example, TASW may be first appraised as a threat, but after evaluating sufficient coping mechanism (such as problem- or emotion-focused coping, see details in Sect. 2.5) it can be re-appraised into a challenge.

In general, if TASW will be appraised as a stressor, resource, or demand may depend on state recovery [Path 3, Fig. 5.1] as well as on individual and environmental factors [Path 4, Fig. 5.1] which are noted below. After work before showing TASW behavior, the appraising may primarily depend on previous experiences with TASW. Further, the anticipated consequences of TASW may also play a substantial role. Furthermore, we expect that appraisals of TASW moderate the relationship between recovery and appraisals as appraising of TASW as resource may diminish the negative relationship between need for recovery and TASW, TASW as stressor may exacerbate it, and TASW as demand may influence it in more or less positive way, depending on its challenging/hindering characteristics [Path 6, Fig. 5.1]. Moreover, depending on appraisals of TASW, we expect differences in the resultant behavior [Path 5, Fig. 5.1]. Employees who appraise TASW as a resource may prefer to do it and those who evaluate TASW as stressor may attempt to avoid it. Employees who judge TASW as a demand may show this behavior more or less, depending on its anticipated challenging/hindering characteristics. The supposed probability of TASW depending on various appraisals is shown in Fig. 5.2.

In sum, as can be seen in Fig. 5.1, we propose that appraisals mediate the relationships between state recovery as well as individual and environmental factors and TASW [Paths 3, 4, 5, Fig. 5.1]. Furthermore, the appraisals may moderate the recovery-TASW relationship [Path 6, Fig. 5.1].

5.1.2   Personal Factors

5.1.2.1   Motivation Behind Work-Related ICT Use After Hours

As we have proposed according to the self-determination theory (SDT) of Deci and Ryan (2000, see Sect. 2.3), employees may be motivated in various ways to show TASW. For example, *intrinsically* motivated employees do it because they enjoy it. *External* motivated employees use ICT for work purposes in their leisure time

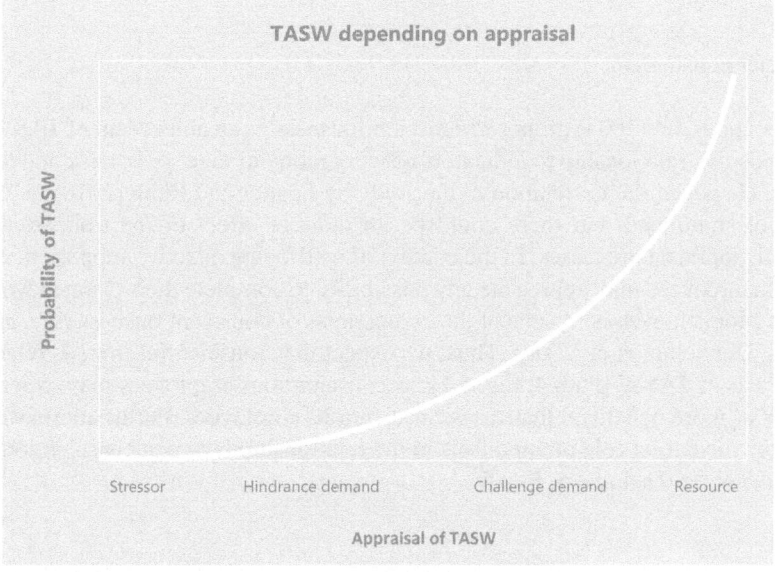

Fig. 5.2 The supposed probability of TASW depending on appraisal

because they expect financial rewards after doing it or in order to avoid being criticized by their colleagues or supervisors after not doing it. *Introjected* motivated employees may use ICT for work purposes in their leisure time in order to obtain feelings such as pride of themselves or to avoid guilt and shame by not doing it. Furthermore, employees *identified* with the importance of supplemental work for their career advancement would use ICT for work purposes in their leisure time more volitionally. At last, *integrated* motivated employees use ICT for work purposes in their leisure time because this behavior fully corresponds with their personal values.

The study by Ohly and Latour (2014) sheds light on the motivation behind TASW as it differentiates between *autonomous* and *controlled* motivation for TASW. The results provide moderate correlations for autonomous as well as for controlled motivation and smartphone use. Initially, they suggest that the appearance of TASW behavior (measured with a dichotomous variable) does not depend on the source causing it (external or internal). However, we suppose that there may be an indirect effect of motivations behind TASW on TASW itself through cognitive appraisals [Paths 4 and 5, Fig. 5.1]. In this context, we suppose that the different forms of motivation contribute to developing various forms of TASW-appraisal. For example, autonomous motivation for TASW may be stronger related to appraising TASW as a resource than as a demand. Further, building on SDT and the findings of Ohly and Latour (2014), we expect various effects of TASW on recovery and well-being in dependence on motivation for TASW. These are discussed in the Sect. 5.3.3.

5.1.2.2   Traits

Conscientiousness

Fenner and Renn (2004) proposed conscientiousness as an antecedent of TASW due to its positive relationship to indicators of work motivation (e.g., Barrick and Mount 1991). However, the correlation in the study by Fenner and Renn (2010) to TASW was not significant, but there could be an indirect effect of the trait on TASW through appraisal processes. In the context of well-being, also we suppose that conscious employees may appreciate any possibility to complete their planned work, as striving for achievement is one of the components of conscientiousness (see, among others, Donnellan et al. 2006). Thus, we expect that conscientiousness is related to appraisals of TASW [Path 4, Fig. 5.1], as conscientious employees may experience TASW as more beneficial than non-conscientious employees. Furthermore, we propose the mediating role of appraisals in the relationship between conscientiousness and TASW [Paths 4 and 5, Fig. 5.1].

Polychronicity

Polychronicity refers to an ability and preference to do many tasks simultaneously (e.g., Hall 1959). Richardson and Benbunan-Fich (2011) expect polychronic individuals to overlap work and non-work time and therefore to be connected to work during after-hours. Their results suggest that polychronicity is positively associated with TASW. Consequently, we expect that polychronicity is positively related to appraisals of TASW, which this, in turn, may influence showing TASW [Paths 4 and 5, Fig. 5.1].

Role Segmentation–Integration

According to the work-family border theory (Clark 2000) and boundary theory (Ashforth et al. 2000), employees differ in the creation, maintenance, and modification of boundaries between work and home domain (see Sect. 2.1). These differences depend, for example, on the extent of identification with the domain roles (employee role vs. family member role). Some evidence shows that highly identified roles will be integrated into other domains by increases in work to non-work permeability and role-referencing (Olson-Buchanan and Boswell 2006).

In this context, the previous study of Richardson and Benbunan-Fich (2011) reveals positive correlation between role segmentation–integration preference and TASW. Therefore, we expect that employees who prefer integrating (labeled integrators) of their work and family roles are more likely, whereas those who prefer segmenting (labeled segmentors) are less likely to appraise TASW as beneficial, and, in turn, to show TASW behavior [Paths 4 and 5, Fig. 5.1]. In the case of segmentation preference, we would expect demanding or stressing, whereas by integration preference, we would expect resourcing associations of TASW.

5.1.2.3 Attitudes Towards Work and Organization

Career Ambition

Ambition, or career commitment, refers to aspiration toward career-related goals (Judge and Kammeyer-Mueller 2012). The self-determination theory (SDT) of Deci and Ryan (2000, see Sect. 2.3) posits that people who identify with the values accept them as a part of their self rather than by introjection. Therefore, we assume that employees with higher career commitment and ambition will use ICT for work purposes in their leisure time more because they do so rather voluntarily. In previous research, Fenner and Renn (2004) proposed career ambition as an antecedent of TASW. Further, Boswell and Olson-Buchanan (2007) have already supported the assumption that employees with higher ambition are more likely to show TASW behavior. This is also shown by the following example from a female middle manager:

I think if I had a family my feelings towards working unsociable hours or checking e-mails would be different. However, as I'm young and wanting to get on in my career I think it is necessary to put in the additional time and effort. (Waller and Ragsdell 2012, p. 168)

In our context, we propose the mediating role of appraisals between career ambition and TASW [Paths 4 and 5, Fig. 5.1]. Thus, we expect more positive appraisals of TASW for high ambitious employees because TASW may be seen as a means to achieve their career-related goals [Path 4, Fig. 5.1].

Affective Organizational Commitment

Another concept which refers to individuals' identification, but in the context of organization, is the affective attachment to an organization, labeled also as affective commitment. Because the employees who identify with the organization may also more likely show TASW behavior voluntarily, we expect similar associations to TASW for affective organizational commitment as for career ambition. Thus, affective commitment will be associated with more positive appraisal of TASW [Path 4, Fig. 5.1]. Previously, Boswell and Olson-Buchanan (2007) assumed that affective commitment is related to supplemental work and therefore to TASW. As a result, the positive association between affective commitment and TASW did not reveal significance. Again, we propose the mediating role of appraisals [Paths 4 and 5, Fig. 5.1].

Work Role Identification, Work Centrality, and Job Involvement

As noted above, one of the potential antecedents/correlates of integration versus segmentation preference is role identification (Clark 2000, see Sect. 2.1). Therefore, we expect that the greater the work role identification, the more likely employees may prefer flexible and permeable boundaries between work- and family roles, and the more positively they may appraise the opportunity of TASW [Path 4, Fig. 5.1].

Based on SDT by Deci and Ryan (2000, see Sect. 2.3), *integrated regulated behavior* represents the most complete and effective internalized form of extrinsic motivation. The resulting behavior is autonomous and fully volitional, although still extrinsic. Therefore, we assume that integrated motivated employees (high on work centrality) appraise working in their leisure time more positively because this behavior fully corresponds with their personal values [Path 4, Fig. 5.1]. This, in turn, may influence TASW [Path 5, Fig. 5.1].

In previous research, Fenner and Renn (2004) proposed positive relationship between job involvement and TASW. In fact, Boswell and Olson-Buchanan (2007) and Park and Jex (2011) showed significant correlations. We propose the mediating role of appraisals in this relationship [Paths 4 and 5, Fig. 5.1].

As an additional construct that is not an attitude but one of the compulsive behavioral tendencies, we assume a positive relationship between *workaholism* and appraising TASW as a demand [Path 4, Fig. 5.1]. In contrast to employees who are attached to their organization, identify with their work, integrate work in their self-concept and therefore work after hours because they enjoy it, workaholics work because they feel a compulsion do it (Schaufeli et al. 2008).

(Work) Control Aspiration

People attempt to gain control over their environment and avoid uncontrollability (see, e.g., the learned helplessness model by Seligman 1975 or two-process model by Rothbaum et al. 1982), including work environment. In our context, we suppose that employees high on TASW behavior may do it in attempt to increase their control over work environment (e.g., by checking and sending e-mails after hours), such as a project manager expressed:

> I feel like I need to deal with each day's e-mail or otherwise you are just guaranteed to lose or overlook something. If I don't have access to my e-mail, like if I'm ever off campus [out of the office], that's when I feel sort of not in control, like I'm missing something. It's like something could be happening with my project that I'm not aware of. (Barley et al. 2011, p. 898)

Furthermore, employees with a high degree of control aspiration may appraise the possibility of TASW as resource due to the anticipated facilitation of maintaining control. Due to the previous, substantial role of control aspiration for important outcomes of organizational well-being (such as personal initiative, see, among others, Frese et al. 2007), we posit its substantial role also in our conceptual framework. We propose the same effects of control aspiration as for other attitudes towards work and organization listed above [Paths 4 and 5, Fig. 5.1].

5.1.2.4   Attitudes Towards ICT

Many studies support the proposition that attitudes toward technology influence its adoption and use. For example, Fenner and Renn (2010) showed that TASW is positively related to perceived usefulness of ICT. Furthermore, according to Diaz et al. (2012), the greater the perceived ICT flexibility, the more likely employees use ICT to perform their job during non-work hours. Most of the studies are based on the technology acceptance model (TAM; Davis 1986) and the unified theory of acceptance and use of technology (UTAUT; Venkatesh et al. 2003). The two models are well established in the previous literature. Therefore, we assume that attitudes towards ICT play a role in showing TASW behavior. We expect that positive attitudes towards ICT increase the positive appraising of TASW by employees, while emphasizing its resourcing functions (see also COR theory; Hobfoll 1989, 2011 in Sect. 2.4) [Path 4, Fig. 5.1]. This, in turn, may increase the likelihood of TASW [Path 5, Fig. 5.1].

5.1.2.5   Habits

In addition to the antecedents discussed above, the action theory points out that not all actions are consciously executed (see Sect. 2.8). Rather, when a behavior is well-practiced, a flexible action pattern (or habits) develops that is executed each time a relevant signal occurs (see also Ouellette and Wood 1998). In the case of ICT, it is likely that checking e-mails or responding as soon as possible is regulated on this level of flexible action patterns, because this behavior is repeated often and under stable circumstances. Relevant signals include the flashing of the phone or the symbol indicating new e-mail. The idea of TASW as a habit is also evident in research assessing ICT use with items such as "When my smartphone blinks to indicate new messages, I cannot resist checking them" (Derks and Bakker 2014). This item reflects that it is not always a conscious decision to check e-mails or to respond to them. Furthermore, the role of habits has also been integrated in the extended unified theory of acceptance and use of technology (UTAUT2; Venkatesh et al. 2012). Therefore, we propose that TASW becomes a habit the more often this behavior is shown (see also Kim and Malhotra 2005; Lankton et al. 2010; Oulasvirta et al. 2012).

Future research could explore this idea further, and assess the degree to which TASW is consciously executed or part of a habit. Moreover, it should be examined, how the extent of habitual TASW may influence its appraising [Path 4, Fig. 5.1]. For example, we would assume that habitual TASW is generally associated with more positive appraisals than non-habitual TASW as we know, that habits are associated with low negative affect (Wood et al. 2002). However, a specific incident of TASW can also be appraised as threatening when it deviates from the habitual TASW. Thus, more research on the beneficial or detrimental consequences of habitual TASW is needed [Path 5, Fig. 5.1]. Furthermore, habitual TASW should be conceptually distinguished from compulsive behavior (for details on technology addiction and its effects see, e.g., Magsamen-Conrad et al. 2014; Salanova et al. 2013; Turel et al. 2011), which is described by a spouse of one of the partners at a small private equity group in following example:

> It is eternal. That is a big word, but I think it's accurate. It is addictive and it never goes away. It's right there. It's easy to use. It's expected. It's a Crackberry, that's the way it is. It's just like crack. (Mazmanian et al. 2013, p. 1347)

In the context of habitual supplemental work behavior via ICT, we further assume a relevant role of *trait self-control* (see, e.g., Baumeister et al. 2006) in forming habits (see, e.g., de Ridder et al. 2012) and therefore also in showing TASW. Some examples from the qualitative research have already pointed the importance of self-control for TASW:

> I know that the technology does make it difficult for some people. . . but it can work as long as you have that self-discipline. (Duxbury et al. 2014, p. 579)
>
> People complain that the BlackBerry ties them to the office. They call it an electronic leash. From my perspective people need to take responsibility for their own actions. Just because I have a BlackBerry doesn't mean I am on call 24 hours a day. If someone sends me an email at 10 o'clock at night it doesn't mean I will be seeing it at 10 o'clock. (Duxbury et al. 2014, p. 579)

Furthermore Al-Dabbagh et al. (2014) introduced the construct of ICT self-discipline as "an individual's ability to control their behaviors towards use of ICTs" (see also Soror et al. 2015). Due to the scarcity of research in this specific field, we propose trait self-control as an additional construct in our conceptual model and call for further research about its effects.

5.1.3   Environmental Factors

5.1.3.1   Organizational Factors

Organizational Culture

Organizational culture refers to the social norms and expectations about appropriate behavior. In the case of TASW, the organizational culture comprises norms and expectations about how and when to use ICT in the evenings. Fenner and Renn (2010) tested the impact of organizational expectations (as a facet of organizational culture) on TASW. To illustrate this concept, an example is given: One item of organizational expectations was: "How would you rate your employer's efforts to measure and track employees' use of technological tools to work from their homes at night or on weekends?" (p. 69). In fact, they found a positive relationship between the extent of expectations and showing TASW behavior. Similarly, a study by Richardson and Benbunan-Fich (2011) confirmed the assumption that TASW is significantly related to the organizational distribution of wireless devices (which may be understood as an implicit request to TASW by employees) and organizational norms about TASW perceived by employees. The distribution of devices may signal that the organization expects employees to extend their work day by using them, such as a sales representative has already mentioned before receiving a BlackBerry:

> If they gave me a BlackBerry, they would want improved communication, which means that they want you to respond to messages quicker.... Honestly, I'm not interested. (Mazmanian 2013, p. 1225)

As it has been noted that organizational culture refers to social norms and expectations about appropriate behavior, at this point it should be mentioned that social norms and expectations are made by people. Thus, in the following we concentrate on the role of people inside organizations which may influence the TASW behavior of employees. Based on social learning theory (Bandura 1965, 1977, 1986; Bandura and Walters 1963, see Sect. 2.2), employees learn by imitating their important social referents, especially when they see that certain behaviors are rewarded. In the work context, some role models, such as supervisors and colleagues, are more appropriate for employees to imitate than others. In particular, supervisors make potent models, because employees perceive them as respected and authoritative figures. Therefore, we expect that if supervisors use ICT for work purposes during after-hours, it will have an impact on the attitudes and behavior of their employees toward TASW. Thus, these employees will show ICT-related supplemental work behavior in order to increase their chances of obtaining positive reinforcement from their supervisors.

Furthermore, according to the theory, behaviors of important social referents (e.g., such colleagues) with whom employees can easily identify are likely to be copied. In particular, we suppose that the likelihood of employees' ICT use after hours will be higher after observing colleagues' ICT-related supplemental behavior (e.g., reading and sending e-mails after hours, taking laptop and work home), and it will be even more enhanced and positively appraised after observing colleagues' reinforcement for it (e.g., acknowledgment from the supervisor or clients). In this context, the study of Park et al. (2011) confirmed the perceived segmentation norm as a predictor of TASW supporting the role of social norms. There is also evidence from qualitative research for the assumption of social norms on own perceptions:

> *They were in competition with each other to see who could cover the most e-mails in a day ... if someone went home and switched off their BlackBerry at say 8 p.m., then they switched on the next morning and found out that all their colleagues had been e-mailing each other and doing all this stuff, they individually perceived that they would be seen to be not working as hard and not as dedicated as these people who had been working up until midnight sending e-mails* (MacCormick et al. 2012, emphasis in original)

Furthermore, interviews by Barley et al. (2011) show that some employees have already recognized their own responsibility by creating and perpetuating the norm of availability respectively responsiveness. For instance, one escalation engineer explained:

> I think different people have slightly different expectations of when you read e-mails. I think it is based on your previous levels of response. I typically respond within reasonable time scales and, therefore, people have that expectation of me. But the flipside to that is that I expect people also to respond to my e-mail within reasonable time scales. (Barley et al. 2011, p. 899)

Taken together, we suppose that employees' TASW behavior is similar to their perception of ICT use of their important social referents (e.g., supervisors and colleagues). Thus, we expect that the individual norms and attitudes toward ICT use become more similar in the course of time to those of their organization, department, work group, or any other significant social groups. Therefore, perceived organizational norms may influence the appraising of TASW as stressor, demand, or resource [Path 4, Fig. 5.1], which, in turn, may have an impact of showing TASW [Path 5, Fig. 5.1].

Organizational Climate

Glavin and Schieman (2010) underscored the aspects of interpersonal work contexts as they showed that both supportive (workplaces where participants experienced social support) as well as conflictive work contexts (workplaces where participants experienced interpersonal conflict) are related to more TASW. However, their results also indicated that employees appraised TASW as detrimental or beneficial for their work and family life, depending on organizational climate.

Thus, organizational climate may influence the appraising of TASW [Path 4, Fig. 5.1], which, in turn, may lead to various TASW behavior [Path 5, Fig. 5.1].

5.1.3.2 Work Characteristics

Task Characteristics

Schieman and Glavin (2008) examined the effects of schedule control and job autonomy on TASW. The positive relationship between schedule control and TASW was stronger among men, whereas the effect of job autonomy was not dependent on gender. Glavin and Schieman (2012) found that job authority, schedule control, and decision-making latitude were positive related to work-family role blurring. Furthermore, Senarathne Tennakoon et al. (2013) reported positive relationship to work flexibility. Conversely, Lanaj et al. (2014) did not find significant correlation between job control and TASW. The expected positive outcomes of job, respectively task, characteristics are in accordance to the well-established job characteristics model (Hackman and Oldham 1980) that proposes the motivational potential of job resources. Due to the specificity of our main topic, we assume task autonomy, task significance/importance, and task urgency–in regard to the *anticipated tasks after hours*–to be substantial predictors of TASW-appraisals [Path 4, Fig. 5.1].

5.1.3.3 Non-Work Characteristics

Work to Non-Work Permeability

According to border theory (Clark 2000, see Sect. 2.1), not just the individual preference is constitutive for the actual integrating or segmenting behavior. The characteristics of borders can influence it as well. We assume that the non-work permeability influences the appraisals of TASW. For example, TASW is not easily shown if the borders are strong–impermeable, inflexible and not allowing blending–; such as when a spouse insists that the person does not work from home in the evenings. This may lead to appraising TASW as more demanding/stressing, and, in turn, to avoiding it [Paths 4 and 5, Fig. 5.1].

Non-Work Demands

Consistent with assumptions concerning work domain, we propose a relationship between non-work demands and TASW through appraisals of TASW. Thus, employees with high non-work demands, such as parents' duties or caring for dependents, may appraise TASW more negatively [Path 4, Fig. 5.1], which, in turn, may lead to decreasing TASW [Path 5, Fig. 5.1] (See also Venkatesh and Vitalari 1992).

Non-Work Culture

As noted above, according to the social learning theory (Bandura 1965, 1977, 1986; Bandura and Walters 1963, see Sect. 2.1), people tend to imitate behaviors and adopt behavioral norms of important social referents and not only in the work context. Thus, employees who perceive a strong norm for work-home segmentation by their family and friends will be likely to create stronger boundaries between their work and non-work roles and, therefore, appraise TASW more negatively [Path 4, Fig. 5.1]. This, in turn, may lead to avoiding TASW [Path 5, Fig. 5.1].

ICT Characteristics

Specific characteristics of ICT, such as ICT overall performance, configurability, portability, or internet connectivity may influence the appraisals of TASW (see e.g., Yun et al. 2012). This can be explained with JD-R model (Bakker and Demerouti 2007; Demerouti et al. 2001) as well as with the transactional model of stress (Lazarus and Folkman 1984) as can be seen in the review by Day et al. (2010). In our context, we would expect, for example, that anticipated ICT hassles may cause appraising of TASW as a stressor or hindrance demand [Path 4, Fig. 5.1], and therefore TASW would be less preferred [Path 5, Fig. 5.1].

Figure 5.3 summarizes the proposed antecedents of TASW.

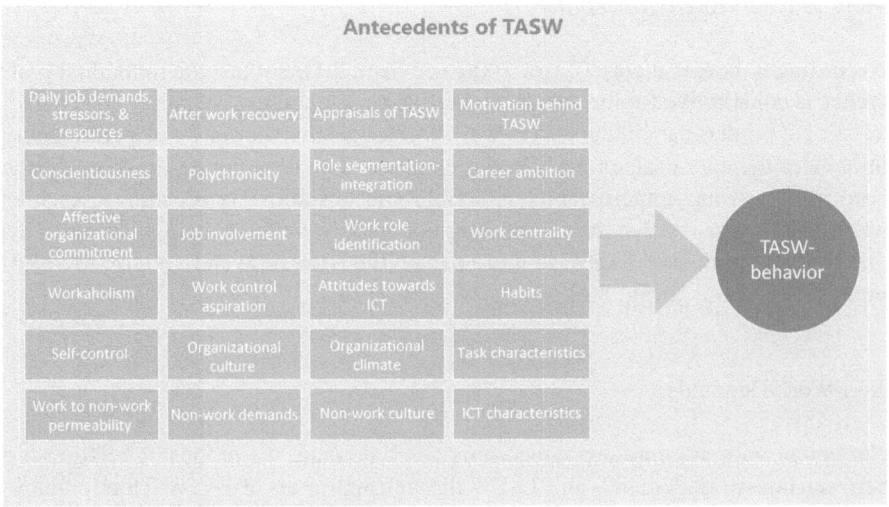

Fig. 5.3 Potential antecedents of TASW

5.2 Consequences of Work-Related ICT Use After Hours

In this section, we combine the mentioned theories and empirical results in regard to consequences of TASW with the emphasis on daily recovery and well-being processes. We consider only working days as we expect different patterns of relationships during the weekend or on holiday due to, for example, absence of scheduled work. Figure 5.4 presents the proposed relationships.

5.2.1 Consequences in the Evening

According to the effort-recovery model (Meijman and Mulder 1998, see Sect. 3.1.1), we propose that TASW is negatively related to evening *recovery* as it drains the same resources (like self-control; see Baumeister et al. 1998; Muraven and Baumeister 2000, see Sect. 3.1.1) as during the scheduled working time and as it hinders the reduction of acute load reactions due to effort expenditure by working [Path 7, Fig. 5.4]. Derks et al. (2014) and Park et al. (2011) have already confirmed this hypothesis as they found TASW negatively related to psychological detachment. In regard to the outcomes of evening recovery, it is associated with indicators of well-being (e.g., Newman et al. 2014; Richardson and Thompson 2012; Sonnentag and Bayer 2005; Sonnentag and Fritz 2007) [Path 8, Fig. 5.4] and

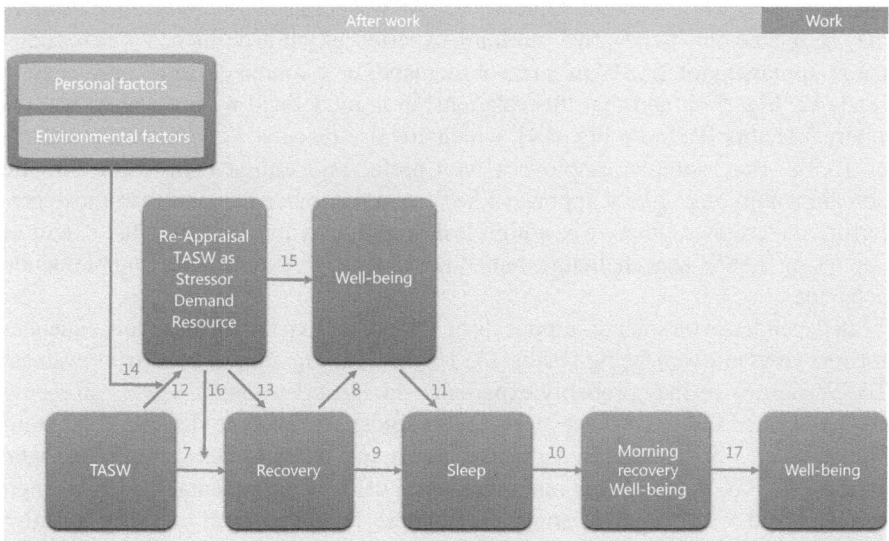

Fig. 5.4 The conceptual model of consequences of TASW with the focus on recovery and well-being processes

following sufficient sleep (e.g., Sonnentag and Fritz 2007) [Path 9, Fig. 5.4] which are important precursors of morning recovery and well-being indicators (see for a review Demerouti et al. 2009) [Path 10, Fig. 5.4]. In this context, there is initial, empirical evidence of the mediating effect of TASW on sleep through evening recovery indicators, such as psychological detachment (see Barber and Jenkins 2014) [Paths 7 and 9, Fig. 5.4]. Furthermore, well-being in the evening may also positively affect sleep (see, e.g., Sonnentag and Fritz 2007) [Path 11, Fig. 5.4].

When considering the *well-being* outcomes of TASW, various (un-)desirable outcomes are expected. Compatible with the 'double-edge sword hypothesis', Diaz et al. (2012) found a positive relationship to work satisfaction as well as to work-life conflict. This leads us back to the assumption of an important role of third variables, like cognition and/or personal as well as environmental factors in the TASW-recovery/well-being process. These third variables might determine if TASW leads to positive or negative outcomes. Thus, we expect that the association between TASW and both level of evening recovery and well-being will be mediated by *re-appraisals* of TASW after doing it [Paths 12, 13, 15, Fig. 5.4]. According to Lazarus and Folkman (1984), human beings evaluate continuously every event when facing it. Thus, they do so before they act, during the action, as well as upon completion of the act (see Sect. 2.5). Similarly, as noted above in the section about antecedents of TASW, employees may evaluate TASW after its completion as a *resource* (such as a means of goal attainment), *stressor* (loss or harm), or *demand* (threat or challenge). Furthermore, in combination with the challenge-hindrance framework (Cavanaugh et al. 2000; LePine et al. 2005), we posited a finer differentiation of demands–as challenge demands and hindrance demands. Thereby, TASW may be appraised as a *challenge demand* after experiencing its strain as well as resourcing characteristics while doing it. Otherwise, TASW appraised as a *hindrance demand* may be related only to negative, straining experiences while doing it. We expect that the re-appraising of TASW as stressor, demand, or resource depends on its extent [Path 12, Fig. 5.4], and that this relationship is moderated by internal as well as external factors [Path 14, Fig. 5.4], which are also discussed above as antecedents of TASW. For example, employees who prefer segmenting their work and life domain would more likely appraise TASW as a hindrance demand than those preferring integrating. Otherwise, a high task importance may enhance the extent of appraising TASW as a challenge demand even by high extent of this supplemental behavior.

In dependence on such re-appraisals of TASW, we expect different consequences for recovery and well-being [Paths 13, 15, and 16, Fig. 5.4]. Those who evaluate TASW as *stressor* may probably experience the most detrimental effects on recovery (evidenced in higher level of glucocorticoids; see Gaab et al. 2005) and well-being (such as higher negative affects) after doing it. Conversely, employees who appraise TASW as a *resource* may experience the least detrimental effects on their recovery and well-being. In some cases, there might be even expected positive effects on specific outcomes, such as job satisfaction (see Diaz et al. 2012). This is in accordance with the assumption of COR theory (Hobfoll 1989, 2011, see Sect. 2.4), that resources create other resources. Furthermore, employees who judge

TASW as a *demand* may experience its detrimental effects on recovery and well-being depending on whether it is appraised as challenging or hindering. Thus, TASW evaluated as a mastery experience may serve as *a challenge demand* and, therefore, lead to less detrimental effects on overall recovery and well-being (see, e.g. Newman et al. 2014). In extreme positive cases, the outcomes might be comparable to the outcomes of appraising TASW as a resource. On the other hand, there may be TASW experienced, for example, as a source of role ambiguity, and therefore as a *hindrance demand* leading to more detrimental effects on overall recovery and well-being; in extreme negative cases comparable to the outcomes of appraising TASW as a stressor. The assumption is also in accordance with JD-R model (Bakker and Demerouti 2007; Demerouti et al. 2001, see Sect. 2.6). The specific outcomes of the extreme cases, when demands turn into stressors, will be further discussed in the section below which presents the final research model.

Moreover, we propose moderating roles of TASW appraisals in the TASW-recovery relationship [Path 16, Fig. 5.4]. Thus, we expect that appraising TASW as a resource may diminish the negative relationship between TASW and recovery, TASW as a stressor may exacerbate it, and TASW as a demand may influence it depending on its challenging/hindering characteristics. Further examples are presented in Sect. 5.3.

5.2.2   Consequences Next Working Day

Thus, as we noted above, sufficient sleep is an important precursor of morning recovery and (organizational) well-being indicators (e.g., Barnes 2012; Sonnentag et al. 2008) [Path 10, Fig. 5.4]. Lanaj et al. (2014) showed that smartphone use for work at night (after 9 pm) reduced sleep quantity, which, in turn, increased morning depletion, and this process had negative consequences for daily work engagement. In accordance with previous research (e.g., Punamäki et al. 2007), we propose positive relationship between sleep and daily well-being which is mediated by morning recovery and well-being [Paths 10 and 17, Fig. 5.4]. This is based on the basic assumptions in recovery research (see, among others, effort-recovery model; Meijman and Mulder 1998; or Chap. 3 in this work).

5.3   Conceptual Framework of Work-Related ICT Use After Hours, Recovery, and Well-Being

As can be seen in the present chapter, the TASW phenomenon is highly complex, as it can be linked to many various antecedents and complex causal chain of consequences for employees' recovery and well-being. Figure 5.5 provides our overall conceptual framework of TASW, recovery and well-being process on working days.

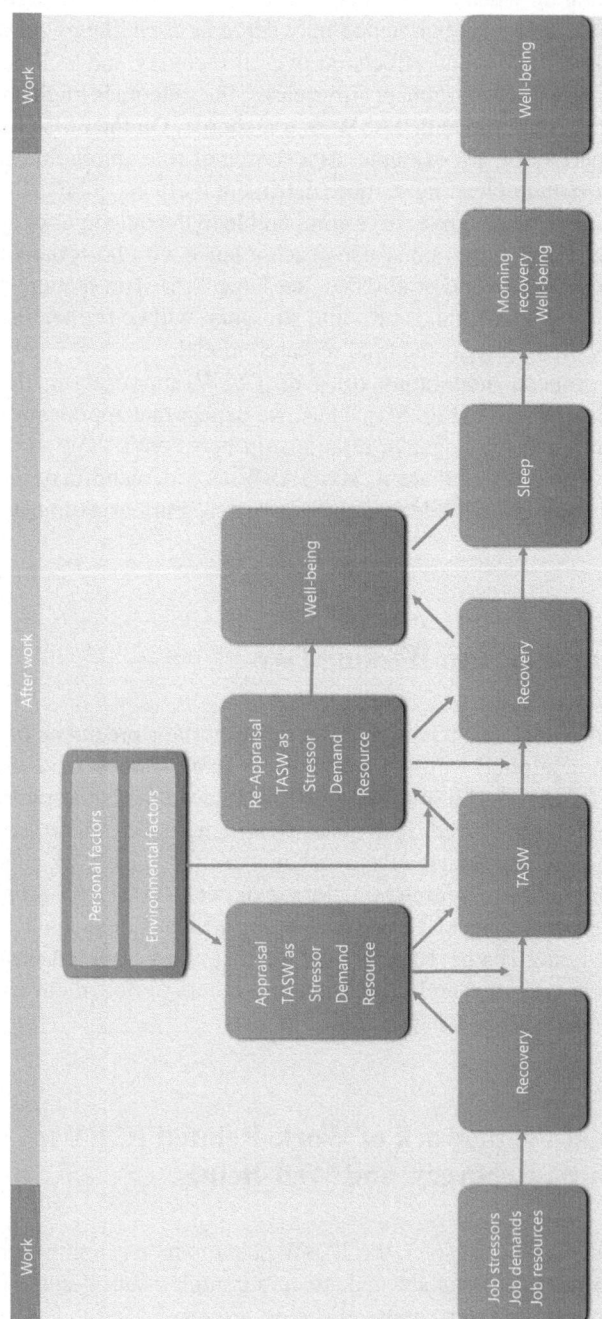

Fig. 5.5 The conceptual framework of TASW, recovery, and well-being processes on working days

As we outlined at the beginning this chapter, we propose TASW to be a potential stressor, resource, and/or demand.

We explain further that it can be evaluated before (described above as appraisal) or directly after doing it (described above as re-appraisal). At this point, we assume that these appraising processes may continue as long as the consequences of TASW persist or as the anticipation of future TASW emerges. This is in accordance with the transactional model of stress (Lazarus and Folkman 1984) that assumes appraisal processes in every new/potential straining situation faced by an individual. Furthermore, we propose that knowledge workers with a possibility of TASW may develop, over time, an overall subjective anticipation of TASW-consequences depending on their previous experiences and on many personal and environmental variables which are noted above.

Thereby, the major prerequisite for future TASW may be the previous consequences appraised by the individual. As the major efforts of work and organizational psychology focus on predicting perception, experience, and behavior of employees in work and organizational context, in the following we present a detailed view on various consequences of TASW which may be substantial predictors of future TASW. According to Zapf and Semmer (2004), stressors, demands, and resources may influence their outcomes in various ways. Thus, their effects might be, for example, linear negative, linear positive, curvilinear, or buffering in regard to recovery and well-being. Thus, in the following three sections we discuss in detail the cases when TASW may be evaluated as stressor, resource, or demand.

5.3.1   Work-Related ICT Use After Hours as a Stressor

At this point, it should be clarified what makes TASW a stressor. According to action theory, stressors are defined as regulation problems as they disturb the regulation of actions (Frese and Zapf 1994, see Sect. 2.8). In our context, we consider *stressors* as factors that impair the process of recovery and well-being. They will lead to lower recovery and well-being (see, e.g., Dunckel 1985; Zapf and Semmer 2004). A more restricted definition was offered by Demerouti et al. (2001) as they proposed to use the term job stressor "only when an external factor has the potential to exert a negative influence on most people in most situations" (p. 501). Thus, based on the JD-R model (Demerouti et al. 2001, see Sect. 2.6), TASW would be termed as a stressor only when it would lead to decreases in recovery and well-being in most situations and by most employees. Furthermore, according to the transactional model of stress (Lazarus and Folkman 1984, see Sect. 2.5), the terms 'harm' or 'loss' suit the stressor concept more likely. Similarly, COR theory (Hobfoll 1989, 2011, see Sect. 2.4) proposes that loss of resources is always experienced as stressful. Thus, if TASW is directly associated only with losing of resources (e.g., energy level next working day), it may be identified as a stressor in the TASW-recovery/well-being process. Also the failure of return of resource investment (e.g., result of TASW perceived as insufficient) should be regarded as a stressor in accordance to COR theory. In combination of all the mentioned stressor concepts, we propose

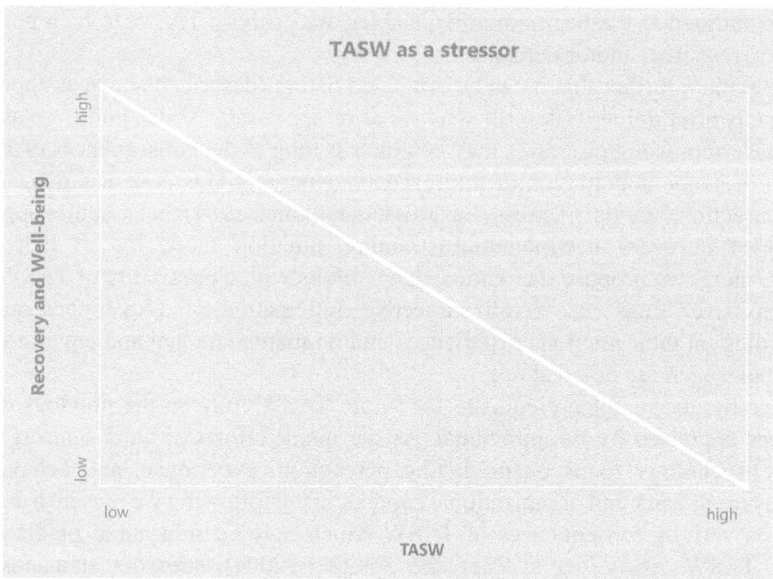

Fig. 5.6 Linear relationship between TASW as a stressor and the core concepts of recovery and well-being

TASW as a stressor in cases when it may be generally assumed to be *linear negative related* to recovery and well-being. Figure 5.6 shows the proposed negative relationship between TASW and our core concepts.

The clearest classification of TASW as a stressor can be provided based on its general negative impact on some *physical outcomes* by the employees. For example, the light-emitting effect of ICT is a topic discussed in the context of recovery and well-being (e.g., Chang et al. 2015; Sroykham and Wongsawat 2013; Wood et al. 2013). As noted above, evening exposure to light-emitting electronic devices extends the time it takes to fall asleep, disrupts the circadian rhythm, suppresses melatonin, modifies REM sleep, and reduces morning alertness (Chang et al. 2015). In this context, Lanaj et al. (2014) have already shown that smartphone use for work at night (after 9 pm) reduced sleep quantity, which, in turn, increased morning depletion, and this process had negative consequences for daily work engagement. Thus, such general detrimental characteristics of ICT make TASW a stressor. Furthermore, the research on *technostress* discusses and examines negative effects of ICT and their use on recovery and well-being (e.g., Ayyagari et al. 2011; Tarafdar et al. 2007; Thomée et al. 2007; Yin et al. 2014). The most general, undesirable effects increasing through use of technological equipment are, for example, eye strain or repetitive strain injuries (such as carpal tunnel syndrome; see, among others, Berolo et al. 2011; Fagarasanu and Kumar 2003; Lai et al. 2014; see also the discussion and research on possible carcinogenicity of mobile phones, e.g., Baan et al. 2011; Roggeveen et al. 2015). Moreover, there are many cases in which *demands* may turn *into stressors*. They are discussed in the section about TASW as a demand below (see Sect. 5.3.3).

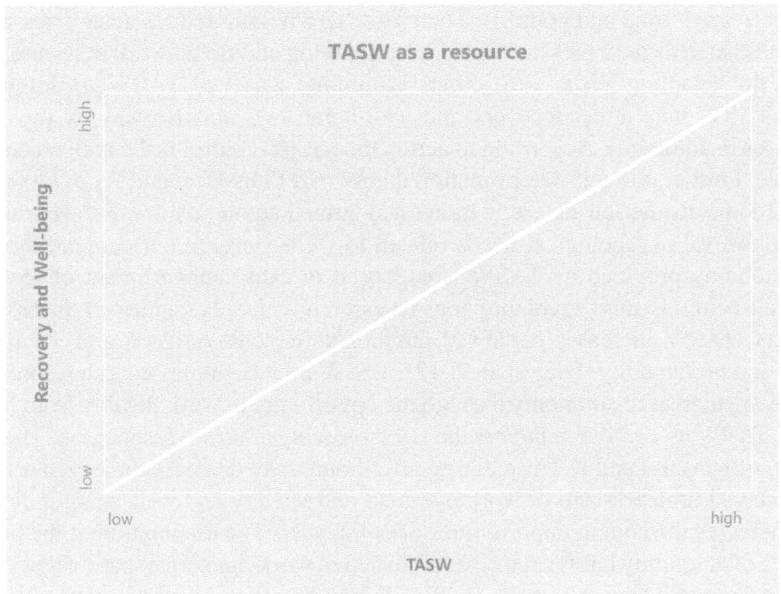

Fig. 5.7 Linear relationship between TASW as a resource and the core concepts of recovery and well-being

5.3.2   *Work-Related ICT Use After Hours as a Resource*

Resources are characterized as means to avoid, hinder, or deal with stressors (Frese 1989; Semmer 1990; Zapf and Semmer 2004). Thus, resources improve recovery and well-being as they reduce the probability of strain occurrence. Furthermore, according to JD-R model (Bakker and Demerouti 2007; Demerouti et al. 2001, see Sect. 2.6), job resources are described as health-protecting aspects of the job which may help by achieving work goals, decrease job demands (as they increase the coping opportunities of employees in demanding environments such as high workload), and enable personal growth and development. Thus, resources themselves are *positively related* to desirable outcomes. Figure 5.7 shows the proposed positive relationship between TASW and our core concepts.

The factors making TASW a resource need to be clarified. There are three ways how TASW can act as a resource. First, we assume that for some employees (see Sect. 5.1.2) TASW itself may *directly* lead to improvements of recovery and well-being (see Fig. 5.7). Second, TASW may affect recovery and well-being *indirectly* as it *reduces stressors or demands* (e.g., work overload the next day). These stressors or demands, in turn, lose, or at least decrease, their potential negative effects on recovery and well-being. This is in accordance with the COR theory (Hobfoll 1989, 2011, see Sect. 2.4) which proposes that employees put in extra compensatory effort to avoid stress. Thus, they invest existing resources, such as their leisure time, technical equipment, and knowledge in order to deal with actual loss, protect against resource loss (e.g., the perceived self-efficacy under time pressure) or gain new resources (e.g., time and control). Therefore, TASW may reduce acute stressors or demands, such as time pressure at work by enabling additional work at leisure time. The term 'enabling'leads us to other resourcing aspect of TASW–to autonomy. Thus, TASW may affect recovery and well-being in a positive way by providing increases in autonomy. According to action theory, job control is the core resourcing variable. Further, the self-determination theory (SDT) by Deci and Ryan (2000, see Sect. 2.3) posits the self-determination (also termed autonomy or control) as one of three universal psychological needs related to well-being. In our context, the high job autonomy provided by TASW is supposed to cause enhancement of recovery and well-being. In this regard, previous research has already confirmed the assumption that TASW increases perceived autonomy (e.g., Richardson and Thompson 2012) and productivity (Diaz et al. 2012). TASW may be appraised as a resource by perceiving increased autonomy through the opportunity to work flexibly from home. Third, TASW as a resource *buffers* the *stressor/demand-strain* association. The possibility of work-related ICT use during after-hours may decrease the negative effect of perceived time pressure or work overload on recovery and well-being indicators as it increases the coping opportunities of employees. The assumption of the buffering role of autonomy has been well-established in work- and organizational psychology (see, e.g., Spector 2009). Thus, TASW may be experienced by employees as a resource when facing stressors or demands (cf. the autonomy paradox by Mazmanian et al. 2013 and empowerment/enslavement paradox by Jarvenpaa and Lang 2005).

Overall, TASW will be (re-)appraised as a resource when an individual experiences in any way its positive influence on his or her recovery and well-being (as a means of goal attainment or hindering strain). There are three ways how TASW can act as a resource: First, TASW may affect recovery and well-being directly in a positive way. Second, TASW may affect recovery and well-being indirectly as it may reduce stressors. Thirdly, TASW may puffer the stressor-strain association.

5.3.3 Work-Related ICT Use After Hours as a Demand

In research on stress, there are several conceptualizations of demand. In this work, they will be combined to find the most appropriate and wide conceptualization of TASW as a demand. Thus, according to action theory (see Sect. 2.8), demands are "requirements necessary to do a particular task" (Zapf 1993, p. 86). Furthermore, the link between demands and recovery and well-being is supposed to be curvilinear (for empirical support see, e.g., Dunckel 1985; Edwards et al. 1998). Thus, there is an optimal level on demands (according to person-environment fit approach; Edwards et al. 1998; Edwards and Van Harrison 1993), and extreme low or high levels should be detrimental for employees. For example, low levels on task complexity may lead to monotony and boredom, whereas high levels may cause cognitive overload. In the JD-R model (Bakker and Demerouti 2007; Demerouti et al. 2001, see Sect. 2.6), job demands are more broadly defined as physical, social, or organizational job characteristics that require physical or psychical effort or skills and related costs. The greater the sympathetic activation or subjective effort, the greater the physiological costs for the employee. Examples of demands are high work pressure, time pressure, workload, an unfavorable physical environment (e.g., noise, heat), and demanding interactions with clients. The term 'stressor' was rather neglected in previous JD-R research. Some factors that are regarded as demands in JD-R are conceptualized as stressors in our model; particularly, the factors of physical environment, because they most likely have detrimental consequences as discussed above. This is in accordance with the JD-R proposition that a stressor is an external factor that "has the potential to exert a negative influence on most people in most situations" (Demerouti et al. 2001, p. 501). As noted above, in consideration of the challenge-hindrance framework (Cavanaugh et al. 2000; LePine et al. 2005, see Sect. 2.7), we posit some finer differentiation of demands–as challenge demands and hindrance demands. This differentiation is comparable to the concepts of 'challenge' and 'threat' by Lazarus and Folkman (1984). Thus, regarding outcomes of TASW, it may be (re-)appraised as a challenge demand in experiencing its straining as well as resourcing characteristics (such as when high workload and work centrality are present). Interestingly, MacCormick et al. (2012) have already found curvilinear relationship between TASW and work engagement. Otherwise, TASW may be appraised as a hindrance demand when experiencing only negative consequences while doing it (such as by high role conflict and low affective organizational commitment).

According to these stress theories, in our context we conceptualize demands as requirements which cause costs for employees but do not necessarily have negative effects on their recovery and well-being. Thus, the classification of TASW as a demand rather than as a stressor seems to be more subjective and therefore more difficult to generalize. Based on Zapf and Semmer (2004), we assume the *curvilinear relationship* to our core constructs and see this as the key contrast to the concept of stressors. Consequently, in following we propose various curvilinear general relationships between TASW and our core concepts depending on the demand type (challenge or hindrance).

As can be seen in Fig. 5.8, for *challenge demands* we assume the typical inverted U-shaped relationship between TASW and our core concepts. Thus, we expect that TASW facilitates the desirable outcomes upon reaching a certain point. This part of the TASW impact may be even labeled as 'resourcing'. But after that, well-being and recovery may be continuously impaired with increasing TASW. This assumption is in accordance with the person-environment fit approach (Edwards et al. 1998; Edwards and Van Harrison 1993). Furthermore, we propose that experiencing this demand as challenging may depend on the personal and environmental factors, listed above, and their various interactions. For example, highly ambitious employees or employees who experience a temporary high workload may experience an optimal well-being in the evening particularly when they can work after hours as well. However, due to some physiological (necessity to sleep) and psychological limits (e.g., self-regulation capacity), after reaching a certain critical point, further TASW may lead only to impaired well-being as it hinders the fundamentally needed recovery processes. The more resources an employee holds, the more the critical point is expected to shift to the right. Such variations in the non-linear relationship

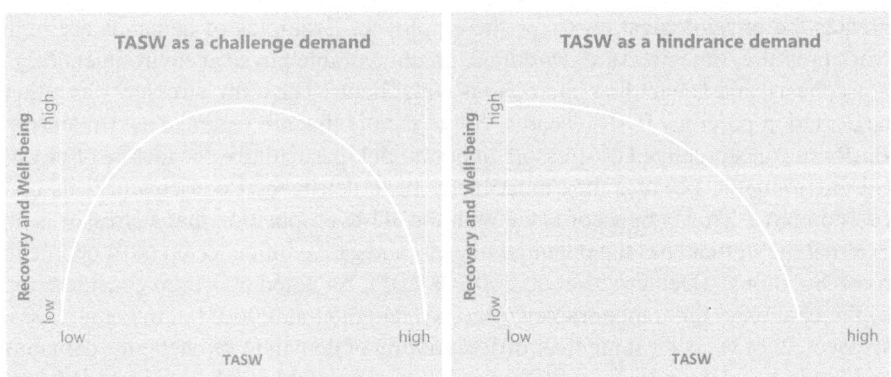

Fig. 5.8 Non-linear relationships between TASW as a challenge- and hindrance demand and the core concepts of recovery and well-being

between TASW as a challenge demand and our core concepts are shown in Fig. 5.9. As seen in Fig. 5.8, for *hindrance demands* we do not assume the typical inverted U-shaped relationship between TASW and our core concepts. We do not expect any positive effects of TASW; we rather expect in the early stages of TASW slow, negative changes and after reaching a certain critical point very quick impairments in well-being and recovery. For example, when employees do TASW only because of external factors (like high task urgency) and in the process they experience TASW as a source of role conflict, they may initially experience only small negative changes in their well-being; however, after reaching a certain critical point, when they do not perceive the amount of TASW as still reasonable, supplemental work anymore, their well-being may drastically decrease. At this point, it should be noted that the critical points after which employees experience drastic impairments of their well-being differ between individuals and depend on available employee resources. The exemplary employee who is externally motivated for TASW (because of time pressure) and who experiences role conflict while doing it, may reach this critical point faster than an employee who has simultaneously strong positive attitude towards work and organization.

The view on the motivation for TASW leads us to another important assumption. According to SDT by Deci and Ryan (2000), we posit that the self-determination continuum (see Fig. 2.3, Sect. 2.3) underlies this resource/demand/stressor relationships. As can be seen in Fig. 5.10, we suggest that autonomy plays a crucial role in the determination of TASW as a resource, demand, or stressor. As long as TASW is determined internally by the individual (intrinsic or integrated motivation), it may have positive effects for the individual. As TASW reaches a certain desirable limit (depending, e.g., on employee characteristics), it will be triggered to some degree by internal- and to some degree by external causes and become a demand. The more external TASW, the more negative consequences it will induce in employees' recovery and well-being. TASW which is caused externally may be harmful already after

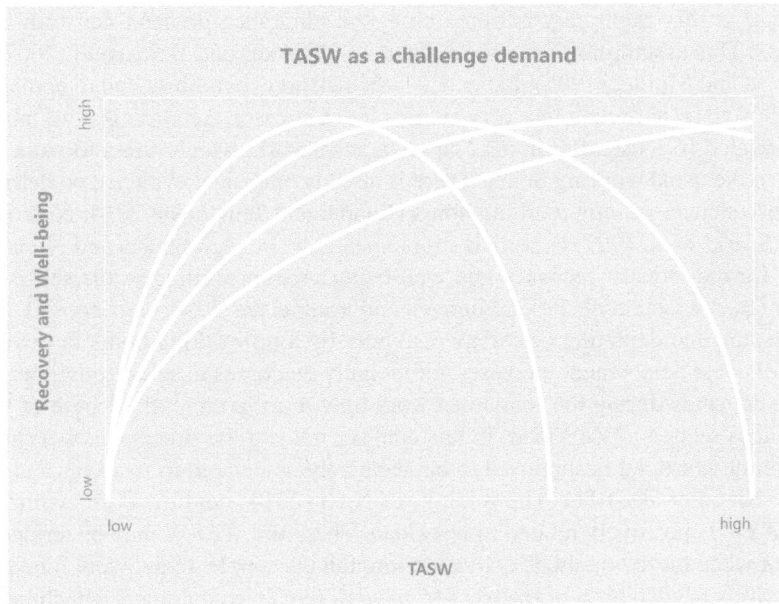

Fig. 5.9 Variations in the non-linear relationship between TASW as a challenge demand and the core concepts of recovery and well-being

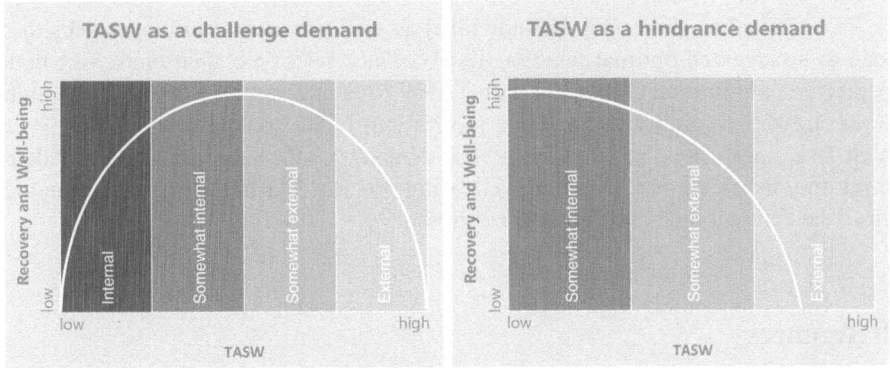

Fig. 5.10 Non-linear relationships between TASW as a challenge- and hindrance demand and the core concepts of recovery and well-being behind the perceived locus of causality on the self-determination continuum

a short period of doing it. Thus, conceivable are various curvilinear relationships which will be flatter the more external the motivation for TASW is. In an extreme case (amotivation or impersonal locus of causality in terms of Deci and Ryan 2000), we would expect a *negative linear relationship* between TASW and well-being and, therefore, determine TASW as a *stressor*.

Thus, in this exemplary extreme case, the hindrance demand can turn into a *stressor*. This assumption is in accordance with Bakker and Demerouti (2007) and Meijman and Mulder (1998). Especially when effort expenditure due to demands is not followed by sufficient recovery, it turns into a stressor. Accordingly, we label the work-related ICT use after hours 'supplemental' work, which refers to work additional to the usual working hours. There is already empirical evidence on detrimental consequences of long working hours (Bannai and Tamakoshi 2014; Nixon et al. 2011; Sparks et al. 1997) as well as supplemental work (Arlinghaus and Nachreiner 2014) for individuals' recovery and well-being. Also according to the self-control model (Baumeister et al. 1998; Muraven and Baumeister 2000, see Sect. 3.1.1), we can assume that depleting of energy resources by long working hours or thwarting psychological detachment increases additionally the losses in self-regulation ability by job demands during the scheduled work time. Furthermore, the length of working time is related to workload. In this context, not just the time spent working but the overall workload is supposed to be negatively associated to recovery and well-being (Nixon et al. 2011). The findings of Nam (2014) confirmed the assumption that TASW is positively related to workload. Therefore, TASW may be termed as a stressor when the sympathetic activation through the supplemental work hinders the subjectively needed level of recovery from work (e.g., psychological detachment) in absence of necessary resources which might buffer this effect (such as flexible work time arrangements that could expand the after-work time). In this case, we assume that TASW will be negatively related to recovery and well-being as it increases the overall working time and hinders recovery.

Taken together, there may be individual as well as external circumstances that lead to a perceived optimal level on TASW. Thus, we expect that there may be a perceived optimum on time spent with TASW for some employees (e.g., with high level on work centrality and internal motivation) under some circumstances (e.g., high task significance and low non-work demands) but exceeding this individual limit diminishes the recovery and well-being process. Furthermore, in an extreme case, the demand may even turn into a stressor.

References

Al-Dabbagh, B., Sylvester, A., & Scornavacca, E. (2014, December 8). *To connect or disconnect – That is the question: ICT self-discipline in the 21st century workplace.* 25th Australasian Conference on Information Systems. ACIS, Auckland, New Zealand.

Arlinghaus, A., & Nachreiner, F. (2014). Health effects of supplemental work from home in the European Union. *Chronobiology International, 31*(10), 1–8. doi:10.3109/07420528.2014.957297.

Ashforth, B. E., Kreiner, G. E., & Fugate, M. (2000). All in a day's work: Boundaries and micro role transitions. *Academy of Management Review, 25*(3), 472–491. doi:10.2307/259305.

Ayyagari, R., Grover, V., & Purvis, R. L. (2011). Technostress: Technological antecedents and implications. *MIS Quarterly, 35*(4), 831–858.

Baan, R., Grosse, Y., Lauby-Secretan, B., El Ghissassi, F., Bouvard, V., Benbrahim-Tallaa, L., et al. (2011). Carcinogenicity of radiofrequency electromagnetic fields. *The Lancet Oncology, 12*(7), 624–626. doi:10.1016/S1470-2045(11)70147-4.

Bakker, A. B., & Demerouti, E. (2007). The Job demands-resources model: State of the art. *Journal of Managerial Psychology, 22*(3), 309–328. doi:10.1108/02683940710733115.

Bandura, A. (1965). Vicarious processes: A case of no-trial learning. In L. Berkowitz (Ed.), *Advances in experimental social psychology* (pp. 1–55). New York: Academic.

Bandura, A. (1977). *Social learning theory*. Englewood Cliffs: Prentice Hall.

Bandura, A. (1986). *Social foundations of thought and action: A social cognitive theory*. Englewood Cliffs: Prentice Hall.

Bandura, A., & Walters, R. H. (1963). *Social learning and personality development*. New York: Holt, Rinehart, Winston.

Bannai, A., & Tamakoshi, A. (2014). The association between long working hours and health: A systematic review of epidemiological evidence. *Scandinavian Journal of Work, Environment & Health, 40*(1), 5–18. doi:10.5271/sjweh.3388.

Barber, L. K., & Jenkins, J. S. (2014). Creating technological boundaries to protect bedtime: Examining work-home boundary management, psychological detachment and sleep. *Stress and Health, 30*(3), 259–264. doi:10.1002/smi.2536.

Barley, S. R., Meyerson, D. E., & Grodal, S. (2011). E-mail as a source and symbol of stress. *Organization Science, 22*(4), 887–906. doi:10.1287/orsc.1100.0573.

Barnes, C. M. (2012). Working in our sleep: Sleep and self-regulation in organizations. *Organizational Psychology Review, 2*(3), 234–257. doi:10.1177/2041386612450181.

Barrick, M. R., & Mount, M. K. (1991). The big five personality dimensions and job performance: A meta-analysis. *Personnel Psychology, 44*(1), 1–26. doi:10.1111/j.1744-6570.1991.tb00688.x.

Baumeister, R. F., Bratslavsky, E., Muraven, M., & Tice, D. M. (1998). Ego depletion: Is the active self a limited resource? *Journal of Personality and Social Psychology, 74*(5), 1252–1265. doi:10.1037/0022-3514.74.5.1252.

Baumeister, R. F., Gailliot, M., DeWall, C. N., & Oaten, M. (2006). Self-regulation and personality: How interventions increase regulatory success, and how depletion moderates the effects of traits on behavior. *Journal of Personality, 74*(6), 1773–1801. doi:10.1111/j.1467-6494.2006.00428.x.

Berolo, S., Wells, R. P., & Amick, B. C. (2011). Musculoskeletal symptoms among mobile hand-held device users and their relationship to device use: A preliminary study in a Canadian university population. *Applied Ergonomics, 42*(2), 371–378. doi:10.1016/j.apergo.2010.08.010.

Boswell, W. R., & Olson-Buchanan, J. B. (2007). The use of communication technologies after hours: The role of work attitudes and work-life conflict. *Journal of Management, 33*(4), 592–610. doi:10.1177/0149206307302552.

Cavanaugh, M. A., Boswell, W. R., Roehling, M. V., & Boudreau, J. W. (2000). An empirical examination of self-reported work stress among U.S. managers. *Journal of Applied Psychology, 85*(1), 65–74. doi:10.1037/0021-9010.85.1.65.

Chang, A.-M., Aeschbach, D., Duffy, J. F., & Czeisler, C. A. (2015). Evening use of light-emitting eReaders negatively affects sleep, circadian timing, and next-morning alertness. *Proceedings of the National Academy of Sciences of the United States of America, 112*(4), 1232–1237. doi:10.1073/pnas.1418490112.

Clark, S. C. (2000). Work/family border theory: A new theory of work/family balance. *Human Relations, 53*(6), 747–770. doi:10.1177/0018726700536001.

Davis, F. D. (1986). *A technology acceptance model for empirically testing new end-user information systems: Theory and results*. Dissertation, MIT Sloan School of Management, Cambridge, MA.

Day, A., Scott, N., & Kelloway, E. K. (2010). Information and communication technology: Implications for job stress and employee well-being. In P. L. Perrewe & D. C. Ganster (Eds.), *New developments in theoretical and conceptual approaches to job stress* (Research in occupational stress and well being, Vol. 8, pp. 317–350). Bingley: Emerald.

de Ridder, D. T. D., Lensvelt-Mulders, G., Finkenauer, C., Stok, F. M., & Baumeister, R. F. (2012). Taking stock of self-control: A meta-analysis of how trait self-control relates to a wide range of behaviors. *Personality and Social Psychology Review, 16*(1), 76–99. doi:10.1177/1088868311418749.

Deci, E. L., & Ryan, R. M. (2000). The "what" and "why" of goal pursuits: human needs and the self-determination of behavior. *Psychological Inquiry, 11*(4), 227–268. doi:10.1207/ S15327965PLI1104_01.

Demerouti, E., Bakker, A. B., Nachreiner, F., & Schaufeli, W. B. (2001). The job demands-resources model of burnout. *Journal of Applied Psychology, 86*(3), 499–512. doi:10.1037/0021-9010.86.3.499.

Demerouti, E., Bakker, A. B., Geurts, S. A., & Taris, T. W. (2009). Daily recovery from work-related effort during non-work time. In S. Sonnentag, P. L. Perrewe, & D. C. Ganster (Eds.), *Current perspectives on job-stress recovery* (Research in Occupational Stress and Well-being, Vol. 7). Bingley: Emerald/JAI Press.

Derks, D., & Bakker, A. B. (2014). Smartphone use, work-home interference, and burnout: A diary study on the role of recovery. *Applied Psychology, 63*(3), 411–440. doi:10.1111/j.1464-0597.2012.00530.x.

Derks, D., van Mierlo, H., & Schmitz, E. B. (2014). A diary study on work-related smartphone use, psychological detachment and exhaustion: Examining the role of the perceived segmentation norm. *Journal of Occupational Health Psychology, 19*(1), 74–84. doi:10.1037/a0035076.

Diaz, I., Chiaburu, D. S., Zimmerman, R. D., & Boswell, W. R. (2012). Communication technology: Pros and cons of constant connection to work. *Journal of Vocational Behavior, 80*(2), 500–508. doi:10.1016/j.jvb.2011.08.007.

Donnellan, M. B., Oswald, F. L., Baird, B. M., & Lucas, R. E. (2006). The Mini-IPIP scales: Tiny-yet-effective measures of the Big Five Factors of personality. *Psychological Assessment, 18*(2), 192–203. doi:10.1037/1040-3590.18.2.192.

Dunckel, H. (1985). *Mehrfachbelastungen am Arbeitsplatz und psychosoziale Gesundheit: Psychologische Überlegungen und aktuarische Analysen.* Frankfurt am Main/New York: P. Lang.

Duxbury, L., Higgins, C., Smart, R., & Stevenson, M. (2014). Mobile technology and boundary permeability. *British Journal of Management, 25*(3), 570–588. doi:10.1111/ 1467-8551.12027.

Edwards, J. R., & Van Harrison, R. (1993). Job demands and worker health: Three-dimensional reexamination of the relationship between person-environment fit and strain. *Journal of Applied Psychology, 78*(4), 628–648.

Edwards, J. R., Caplan, R. D., & Harrison, R. V. (1998). Person-environment fit theory: Conceptual foundations, empirical evidence, and directions for future research. In C. L. Cooper (Ed.), *Theories of organizational stress* (pp. 28–67). Oxford/New York: Oxford University Press.

Fagarasanu, M., & Kumar, S. (2003). Carpal tunnel syndrome due to keyboarding and mouse tasks: A review. *International Journal of Industrial Ergonomics, 31*(2), 119–136. doi:10.1016/ S0169-8141(02)00180-4.

Fenner, G. H., & Renn, R. W. (2004). Technology-assisted supplemental work: Construct definition and a research framework. *Human Resource Management, 43*(2–3), 179–200. doi:10.1002/hrm.20014.

Fenner, G. H., & Renn, R. W. (2010). Technology-assisted supplemental work and work-to-family conflict: The role of instrumentality beliefs, organizational expectations and time management. *Human Relations, 63*(1), 63–82. doi:10.1177/0018726709351064.

Frese, M. (1989). Theoretical models of control and health. In S. L. Sauter, J. J. Hurrell, & C. L. Cooper (Eds.), *Job control and worker health* (pp. 108–128). Chichester/New York: Wiley.

Frese, M., & Zapf, D. (1994). Action as the core of work psychology: A German approach. In M. D. Dunnette, L. M. Hough, & H. C. Triandis (Eds.), *Handbook of industrial and organizational psychology* (Vol. 4, pp. 271–340). Palo Alto: Consulting Psychologists Press.

Frese, M., Garst, H., & Fay, D. (2007). Making things happen: Reciprocal relationships between work characteristics and personal initiative in a four-wave longitudinal structural equation model. *Journal of Applied Psychology, 92*(4), 1084–1102. doi:10.1037/0021-9010.92.4.1084.

Gaab, J., Rohleder, N., Nater, U. M., & Ehlert, U. (2005). Psychological determinants of the corti-sol stress response: The role of anticipatory cognitive appraisal. *Psychoneuroendocrinology, 30*(6), 599–610. doi:10.1016/j.psyneuen.2005.02.001.

Glavin, P., & Schieman, S. (2010). Interpersonal context at work and the frequency appraisal and consequences of boundary spanning demands. *Sociological Quarterly, 51*(2), 205–225. doi:10.1111/j.1533-8525.2010.01169.x.

Glavin, P., & Schieman, S. (2012). Work-family role blurring and work-family conflict: The moderating influence of job resources and job demands. *Work and Occupations, 39*(1), 71–98. doi:10.1177/0730888411406295.

Hacker, W. (1998). *Allgemeine Arbeitspsychologie: Psychische Regulation von Arbeitstätigkeiten.* Bern: H. Huber.

Hacker, W. (2003). Action regulation theory: A practical tool for the design of modern work processes? *European Journal of Work and Organizational Psychology, 12*(2), 105–130. doi:10.1080/13594320344000075.

Hackman, J. R., & Oldham, G. R. (1980). *Work redesign.* Reading: Addison-Wesley.

Hall, E. T. (1959). *The silent language: An anthropologist reveals how we communicate by our manners and behavior* (Anchor books, Vol. 948). Garden City: Anchor Press/Doubleday.

Hobfoll, S. E. (1989). Conservation of resources: A new attempt at conceptualizing stress. *American Psychologist, 44*(3), 513–524. doi:10.1037/0003-066X.44.3.513.

Hobfoll, S. E. (2011). Conservation of resource caravans and engaged settings. *Journal of Occupational and Organizational Psychology, 84*(1), 116–122. doi:10.1111/j.2044-8325.2010.02016.x.

Jarvenpaa, S., & Lang, K. (2005). Managing the paradoxes of mobile technology. *Information Systems Management Journal, 22*(4), 7–23.

Judge, T. A., & Kammeyer-Mueller, J. D. (2012). On the value of aiming high: The causes and consequences of ambition. *Journal of Applied Psychology, 97*(4), 758–775. doi:10.1037/a0028084.

Kim, S. S., & Malhotra, N. K. (2005). A longitudinal model of continued IS use: An integrative view of four mechanisms underlying postadoption Phenomena. *Management Science, 51*(5), 741–755. doi:10.1287/mnsc.1040.0326.

Lai, W. K. C., Chiu, Y. T., & Law, W. S. (2014). The deformation and longitudinal excursion of median nerve during digits movement and wrist extension. *Manual Therapy, 19*(6), 608–613. doi:10.1016/j.math.2014.06.005.

Lanaj, K., Johnson, R. E., & Barnes, C. M. (2014). Beginning the workday yet already depleted? Consequences of late-night smartphone use and sleep. *Organizational Behavior and Human Decision Processes, 124*(1), 11–23. doi:10.1016/j.obhdp.2014.01.001.

Lankton, N. K., Wilson, E. V., & Mao, E. (2010). Antecedents and determinants of information technology habit. *Information & Management, 47*(5–6), 300–307. doi:10.1016/j.im.2010.06.004.

Lazarus, R. S., & Folkman, S. (1984). *Stress, appraisal, and coping.* New York: Springer.

LePine, J. A., Podsakoff, N. P., & LePine, M. A. (2005). A meta-analytic test of the challenge stressor-hindrance stressor framework: An explanation for inconsistent relationships among stressors and performance. *The Academy of Management Journal, 48*(5), 764–775.

MacCormick, J. S., Dery, K., & Kolb, D. G. (2012). Engaged or just connected? Smartphones and employee engagement. *Organizational Dynamics, 41*(3), 194–201. doi:10.1016/j.orgdyn.2012.03.007.

Magsamen-Conrad, K., Billotte-Verhoff, C., & Greene, K. (2014). Technology addiction's contribution to mental wellbeing: The positive effect of online social capital. *Computers in Human Behavior, 40*, 23–30. doi:10.1016/j.chb.2014.07.014.

Mazmanian, M. A. (2013). Avoiding the trap of constant connectivity: When congruent frames allow for heterogeneous practices. *Academy of Management Journal, 56*(5), 1225–1250. doi:10.5465/amj.2010.0787.

Mazmanian, M. A., Orlikowski, W. J., & Yates, J. (2013). The autonomy paradox: The implications of mobile email devices for knowledge professionals. *Organization Science, 24*(5), 1337–1357. doi:10.1287/orsc.1120.0806.

Meijman, T. F., & Mulder, G. (1998). Psychological aspects of workload. In P. J. D. Drenth & H. Thierry (Eds.), *Handbook of work and organizational psychology* (pp. 5–33). Hove: Psychology Press.

Muraven, M., & Baumeister, R. F. (2000). Self-regulation and depletion of limited resources: Does self-control resemble a muscle? *Psychological Bulletin, 126*(2), 247–259. doi:10.1037/0033-2909.126.2.247.

Nam, T. (2014). Technology use and work-life balance. *Applied Research in Quality of Life, 9*(4), 1017–1040. doi:10.1007/s11482-013-9283-1.

Newman, D. B., Tay, L., & Diener, E. (2014). Leisure and subjective well-being: A model of psychological mechanisms as mediating factors. *Journal of Happiness Studies, 15*(3), 555–578. doi:10.1007/s10902-013-9435-x.

Nixon, A. E., Mazzola, J. J., Bauer, J., Krueger, J. R., & Spector, P. E. (2011). Can work make you sick? A meta-analysis of the relationships between job stressors and physical symptoms. *Work & Stress, 25*(1), 1–22. doi:10.1080/02678373.2011.569175.

Ohly, S., & Latour, A. (2014). Work-related smartphone use and well-being in the evening: The role of autonomous and controlled motivation. *Journal of Personnel Psychology, 13*(4), 174–183. doi:10.1027/1866-5888/a000114.

Olson-Buchanan, J. B., & Boswell, W. R. (2006). Blurring boundaries: Correlates of integration and segmentation between work and nonwork. *Journal of Vocational Behavior, 68*(3), 432–445. doi:10.1016/j.jvb.2005.10.006.

Ouellette, J. A., & Wood, W. (1998). Habit and intention in everyday life: The multiple processes by which past behavior predicts future behavior. *Psychological Bulletin, 124*(1), 54–74. doi:10.1037//0033-2909.124.1.54.

Oulasvirta, A., Rattenbury, T., Ma, L., & Raita, E. (2012). Habits make smartphone use more pervasive. *Personal and Ubiquitous Computing, 16*(1), 105–114. doi:10.1007/s00779-011-0412-2.

Park, Y., & Jex, S. M. (2011). Work-home boundary management using communication and information technology. *International Journal of Stress Management, 18*(2), 133–152. doi:10.1037/a0022759.

Park, Y., Fritz, C., & Jex, S. M. (2011). Relationships between work-home segmentation and psychological detachment from work: The role of communication technology use at home. *Journal of Occupational Health Psychology, 16*(4), 457–467. doi:10.1037/a0023594.

Punamäki, R.-L., Wallenius, M., Nygård, C.-H., Saarni, L., & Rimpelä, A. (2007). Use of information and communication technology (ICT) and perceived health in adolescence: the role of sleeping habits and waking-time tiredness. *Journal of Adolescence, 30*(4), 569–585. doi:10.1016/j.adolescence.2006.07.004.

Richardson, K., & Benbunan-Fich, R. (2011). Examining the antecedents of work connectivity behavior during non-work time. *Information and Organization, 21*(3), 142–160. doi:10.1016/j.infoandorg.2011.06.002.

Richardson, K. M., & Thompson, C. A. (2012). High tech tethers and work-family conflict: A conservation of resources approach. *Engineering Management Research, 1*(1), 29. doi:10.5539/emr.v1n1p29.

Roggeveen, S., van Os, J., Viechtbauer, W., & Lousberg, R. (2015). EEG changes due to experimentally induced 3G mobile phone radiation. *PLoS ONE, 10*(6), e0129496. doi:10.1371/journal.pone.0129496.

Rothbaum, F., Weisz, J. R., & Snyder, S. S. (1982). Changing the world and changing the self: A two-process model of perceived control. *Journal of Personality and Social Psychology, 42*(1), 5–37. doi:10.1037/0022-3514.42.1.5.

Salanova, M., Llorens, S., & Cifre, E. (2013). The dark side of technologies: Technostress among users of information and communication technologies. *International Journal of Psychology: Journal international de psychologie, 48*(3), 422–436. doi:10.1080/00207594.2012.680460.

Schaufeli, W. B., Taris, T. W., & Bakker, A. B. (2008). It takes two to tango: Workaholism is working excessively and working compulsively. In R. J. Burke & C. L. Cooper (Eds.), *The long work hours culture: Causes, consequences and choices* (pp. 203–225). Bingley: Emerald Group Publishing.

Schieman, S., & Glavin, P. (2008). Trouble at the border?: Gender, flexibility at work, and the work-home interface. *Social Problems, 55*(4), 590–611. doi:10.1525/sp.2008.55.4.590.

Seligman, M. E. (1975). *Helplessness: On depression, development, and death.* San Francisco/ New York: Freeman.

Semmer, N. K. (1990). Stress und Kontrollverlust. In F. Frei & I. Udris (Eds.), *Das Bild der Arbeit* (pp. 190–207). Bern: H. Huber.

Senarathne Tennakoon, K. U., da Silveira, G. J., & Taras, D. G. (2013). Drivers of context-specific ICT use across work and nonwork domains: A boundary theory perspective. *Information and Organization, 23*(2), 107–128. doi:10.1016/j.infoandorg.2013.03.002.

Sonnentag, S., & Bayer, U.-V. (2005). Switching off mentally: Predictors and consequences of psychological detachment from work during off-job time. *Journal of Occupational Health Psychology, 10*(4), 393–414. doi:10.1037/1076-8998.10.4.393.

Sonnentag, S., & Fritz, C. (2007). The recovery experience questionnaire: Development and validation of a measure for assessing recuperation and unwinding from work. *Journal of Occupational Health Psychology, 12*(3), 204–221. doi:10.1037/1076-8998.12.3.204.

Sonnentag, S., Binnewies, C., & Mojza, E. J. (2008). "Did you have a nice evening?" A day-level study on recovery experiences, sleep, and affect. *Journal of Applied Psychology, 93*(3), 674–684. doi:10.1037/0021-9010.93.3.674.

Soror, A. A., Hammer, B. I., Steelman, Z. R., Davis, F. D., & Limayem, M. M. (2015). Good habits gone bad: Explaining negative consequences associated with the use of mobile phones from a dual-systems perspective. *Information Systems Journal, 25*(4), 403–427. doi:10.1111/isj.12065.

Sparks, K., Cooper, C., Fried, Y., & Shirom, A. (1997). The effects of hours of work on health: A meta-analytic review. *Journal of Occupational and Organizational Psychology, 70*(4), 391–408. doi:10.1111/j.2044-8325.1997.tb00656.x.

Spector, P. E. (2009). The role of job control in employee health and well-being. In C. L. Cooper, J. C. Quick, & M. Schabracq (Eds.), *International handbook of work and health psychology* (pp. 171–195). Chichester/Malden: Wiley-Blackwell.

Sroykham, W., & Wongsawat, Y. (2013, March 3). *Effects of LED-backlit computer screen and emotional self-regulation on human melatonin production*, 35th annual international conference of the IEEE EMBS, Osaka, Japan.

Tarafdar, M., Tu, Q., Ragu-Nathan, B., & Ragu-Nathan, T. (2007). The impact of technostress on role stress and productivity. *Journal of Management Information Systems, 24*(1), 301–328. doi:10.2753/MIS0742-1222240109.

Thomée, S., Eklöf, M., Gustafsson, E., Nilsson, R., & Hagberg, M. (2007). Prevalence of perceived stress, symptoms of depression and sleep disturbances in relation to information and communication technology (ICT) use among young adults – An explorative prospective study. *Computers in Human Behavior, 23*(3), 1300–1321. doi:10.1016/j.chb.2004.12.007.

Turel, O., Serenko, A., & Bontis, N. (2011). Family and work-related consequences of addiction to organizational pervasive technologies. *Information & Management, 48*(2–3), 88–95. doi:10.1016/j.im.2011.01.004.

Venkatesh, A., & Vitalari, N. P. (1992). An emerging distributed work arrangement: An investigation of computer-based supplemental work at home. *Management Science, 38*(12), 1687–1706. doi:10.1287/mnsc.38.12.1687.

Venkatesh, V., Morris, M. G., Davis, G. B., & Davis, F. D. (2003). User acceptance of information technology: Toward a unified view. *Management Information Systems Quarterly, 27*(3), 425–478.

Venkatesh, V., Thong, J. Y., & Xu, X. (2012). Consumer acceptance and use of information technology: Extending the unified theory of acceptance and use of technology. *MIS Quarterly, 36*(1), 157–178.

Waller, A. D., & Ragsdell, G. (2012). The impact of e-mail on work-life balance. *Aslib Proceedings, 64*(2), 154–177. doi:10.1108/00012531211215178.

Wood, W., Quinn, J. M., & Kashy, D. A. (2002). Habits in everyday life: Thought, emotion, and action. *Journal of Personality and Social Psychology, 83*(6), 1281–1297. doi:10.1037/0022-3514.83.6.1281.

Wood, B., Rea, M. S., Plitnick, B., & Figueiro, M. G. (2013). Light level and duration of exposure determine the impact of self-luminous tablets on melatonin suppression. *Applied Ergonomics, 44*(2), 237–240. doi:10.1016/j.apergo.2012.07.008.

Yin, P., Davison, R. M., Bian, Y., Wu, J., & Liang, L. (2014). The sources and consequences of mobile technostress in the workplace. In *PACIS 2014 Proceedings. (Paper 144)*. http://aisel.aisnet.org/pacis2014/144

Yun, H., Kettinger, W. J., & Lee, C. C. (2012). A new open door: The smartphone's impact on work-to-life conflict, stress, and resistance. *International Journal of Electronic Commerce, 16*(4), 121–152. doi:10.2753/JEC1086-4415160405.

Zapf, D. (1993). Stress-oriented analysis of computerized office work. *European Work and Organizational Psychologist, 3*(2), 85–100. doi:10.1080/09602009308408580.

Zapf, D., & Semmer, N. K. (2004). Stress und Gesundheit in Organisationen. In H. Schuler (Ed.), *Enzyklopädie der Psychologie*: *Organisationspsychologie – Grundlagen und Personalpsychologie* (Wirtschafts- Organisations- und Arbeitspsychologie, Vol. 3, pp. 1007–1112). Göttingen [u.a.]: Hogrefe.

Chapter 6
Conclusion and Discussion

The aim of this work was to provide insight into the process of employee recovery and well-being in regard to work-related ICT use during after-hours (labelled TASW). Therefore, we discussed firstly theories that have helped us to understand the determinants and outcomes of TASW; secondly, our core concepts recovery and well-being; and, thirdly, previous empirical findings on TASW. Based on literature review, we proposed a novel conceptual, overall framework of TASW with focus on employee recovery and well-being processes. Thereby, we posited TASW as potential stressor, resource, or demand (see action theory by Hacker 1998, 2003; Frese and Zapf 1994), depending on personal and environmental factors, but primarily on cognitive appraisals (see transactional model of stress by Lazarus and Folkman 1984).

In regard to *predicting* the contribution of appraisals for actual behavior, we assumed that employees who appraise TASW as a resource may usually prefer to do it. Those who evaluate TASW as stressor may probably attempt to avoid it. In consideration of the challenge-hindrance framework (Cavanaugh et al. 2000; LePine et al. 2005), we posited some finer differentiation of demands–as challenge demands and hindrance demands. Thus, we assumed that TASW may be appraised as a challenge demand in anticipating its straining as well as resourcing characteristics. Otherwise, TASW may be appraised as a hindrance demand when anticipating only negative experiences while doing it. Consequently, frequency of showing TASW when judging it as a demand may depend on anticipating its challenging or hindering characteristics.

Furthermore, the three-way division (stressor/resource/demand) enabled us to propose various linear and non-linear associations to well-being and recovery *outcomes*. According to established stressor concepts, we posited TASW as a *stressor* in cases when it is linear negative related to recovery and well-being. Further, there are three ways how TASW can act as a *resource*: First, TASW may affect recovery and well-being directly in a positive way. Second, it may affect recovery and well-being indirectly as it may reduce stressors. Thirdly, TASW may puffer the stressor-strain association. The classification of TASW as a *demand* rather than as a stressor seems to be more subjective and therefore more difficult to generalize. Based on Zapf and Semmer (2004), we assumed a curvilinear relationship to our core constructs and see this as the key contrast to the concept of stressors. Consequently, we proposed various curvilinear general relationships between

© The Author(s) 2016
L. Ďuranová, S. Ohly, *Persistent Work-Related Technology Use, Recovery and Well-being Processes*, SpringerBriefs in Psychology,
DOI 10.1007/978-3-319-24759-5_6

TASW and our core concepts depending on the demand type (challenge or hindrance). Thus, there may be individual as well as external circumstances that lead to a perceived optimal level on TASW. In detail, we expect that there may be a perceived optimum on time spent with TASW for some employees under some circumstances but exceeding this individual limit may diminish their recovery and well-being process. Moreover, in an extreme case, the demand may even turn into a stressor.

In the following, we conclude with an overall discussion on further research in regard to the identified research gaps and discuss some future implications for practice. Finally, we present in short the overall contribution of this work.

Due to the scarcity of studies in the focused research field, more well-designed studies based on representative samples (e.g., large samples, variety of professions and industries) are necessary in *future research*–first to reexamine the previous findings, and second to examine further concepts theoretically related to TASW in the context of daily recovery and well-being after work. As we noted above, research focused on investigating the consequences of TASW has greatly increased recently. In contrast, the antecedents remained underexplored, although, in our opinion, previous studies have discussed potentially worthwhile ideas (like attitudes towards technology, see Boswell and Olson-Buchanan 2007; between-person differences as levels of energy, see Derks et al. 2014b; or the not adequately met obligations of the paid work role, such as unfinished tasks, see Glavin et al. 2011). In Chap. 5, we suggested additional factors, which may play an important role in our conceptual model and were not included in the theories or studies mentioned above. These are, for example, workaholism (this should be at least a controlled variable as it is a strong predictor for supplemental work, see Derks et al. 2014a) and non-work demands. Future research could explore the idea that TASW is habitual, and assess the degree to which TASW is consciously executed or part of a habit. This differentiation could have important implications for practice as well, because changing a habit requires specific interventions.

Looking at the consequences of TASW, systematically examining all variables and their various relationships proposed in our model may help to understand the role of TASW in the daily recovery and well-being processes of working people. In particular, the empirical distinction of TASW as a stressor, challenge/hindrance demand, and resource, which depends on appraisals as well as various personal and environmental factors, may be challenging for future research. Thus, also the special assumption of non-linear relationships between TASW as a demand and its outcomes should be examined. In future research, it is advisable to assess general workload and work time to be able to examine the extent to which the effects of TASW are incremental to this established work demand.

As we concluded in Chap. 4, due to the fact that in previous studies TASW was differentially linked with outcomes depending on its operationalization, future research should include multiple operationalizations of TASW, including objective data on use, which could be obtained from smartphone logs. Furthermore, almost all previous studies used self-report data (except Boswell and Olson-Buchanan 2007) and, therefore, might be affected by common method variance (Podsakoff et al.

2003). Thus, the predictors and outcomes of TASW may also be captured more objectively in the future. For example, sleep could be measured not only via self-reports but also through physiological measurements such as sleep actigraphy (Barber and Jenkins 2014) or melatonin production (Lanaj et al. 2014). In addition, other psychological variables, such as rumination about work or organizational norms could be captured via multiple sources (e.g., asking significant others, such as partners colleagues or supervisors).

Most previous studies were cross-sectional in nature and, thus, conclusions about causal effects are premature. Therefore, future research should employ a longitudinal design (Zapf et al. 1996). Due to the short-term nature and, therefore, fluctuations of our core concepts, it can be worthwhile to develop diary designs which enable to test lagged effects (see methodological issues when planning a diary study by Ohly et al. 2010). This design has been employed already by Lanaj et al. (2014).

As already noted by Derks et al. (2015), also a quasi-experimental design in field studies would be conceivable in the context of ICT use. The authors suggest a two-group design with zero-history group on ICT use for business purposes (such as trainees starting their careers). Furthermore, a group with limited ICT use could be compared to a group using ICT permanently. First applications have been developed to help users limit their ICT use, and the evaluation is currently underway (Curtaz et al. 2015; for further ideas, see also David et al. 2014).

In the following, we discuss arguable *implications for practice*. At the current state of research on TASW, it is not possible to posit specific implications for practice (see arguments above, e.g., the lack of longitudinal studies). The state of the art is that TASW may have positive as well as negative outcomes. Therefore, employers should not expect only positive organizational outcomes by supplemental work of their employees after hours, but also consider the costs for an upcoming working day due to impaired recovery and well-being (see also Arlinghaus and Nachreiner 2014). Thus, limiting TASW by discussing the potential of negative consequences, establishing social norms–when and how often one is required to TASW–seem advisable to prevent negative consequences and to reduce telepressure (the urge to respond everywhere and every time, see Barber and Santuzzi 2015 and Sect. 4.1). However, because prohibiting TASW altogether might not be possible and could be perceived as controlling, a balanced approach is needed. Based on the idea that TASW might be a habit, prevention of signals might help employees to reduce TASW and check their e-mail after hours only when truly necessary (e.g., through deactivating the push function on the smartphones or the acoustic signals for incoming messages on devices). In the case of work overload, Dabbish and Kraut (2006) brought evidence that TASW management strategies (like having smaller number of folders and keeping inbox small) serve as resources by evaluating the relationship between work quantity (captured as e-mail volume) and the perceived work overload (captured as e-mail overload). Furthermore, we expect that future examination of our model assumptions would help to develop appropriate intervention strategies depending on various personal and environmental factors. Particularly, it may be essential to know under which circumstances TASW serves as a stressor and should be minimized. For example, it should be clarified which exact role TASW plays for

physiological processes. Not less important is recognizing the role of some resourcing and demanding factors in regard to developing appropriate personnel trainings
(providing information on the pros and cons of TASW and using incentives and
policies) or reorganizing work design (addressing flexible workplace
arrangements).

In sum, the complex assumptions of interdependence between technology use
and person/environment factors is in accordance with sociomaterial view by
Orlikowski (2007). This approach points out that material properties of technology
(e.g., mobility, push functions) as well as the social characteristics (e.g., social
norms, cultural interpretations, individual dispositions) shape technology use and
are in turn influenced by it. Thus, by highlighting the appraisal of TASW and how
the appraisal is dependent on individual and contextual characteristics, this work
contributes to the sociomaterial view to gain knowledge about the conditions that
affect daily recovery and well-being. Given that past research mostly overlooked the
idea that TASW may serve not only as a (challenging/hindering) demand or a
resource but its specific characteristics can turn it into a stressor, this work sheds
light on the varying perceptions of TASW. Furthermore, our final conceptual framework is based on several well-established theories as well as previous initial research
on TASW and, therefore, may be helpful for understanding this under-researched
phenomenon.

References

Arlinghaus, A., & Nachreiner, F. (2014). Health effects of supplemental work from home in the
    European Union. *Chronobiology International, 31*(10), 1–8. doi:10.3109/07420528.2014.957
    297.
Barber, L. K., & Jenkins, J. S. (2014). Creating technological boundaries to protect bedtime:
    Examining work-home boundary management, psychological detachment and sleep. *Stress
    and Health, 30*(3), 259–264. doi:10.1002/smi.2536.
Barber, L. K., & Santuzzi, A. M. (2015). Please respond: Workplace telepressure and employee
    recovery. *Journal of Occupational Health Psychology, 20*, 171–189. doi:10.1037/a0038278.
Boswell, W. R., & Olson-Buchanan, J. B. (2007). The use of communication technologies after
    hours: The role of work attitudes and work-life conflict. *Journal of Management, 33*(4), 592–
    610. doi:10.1177/0149206307302552.
Cavanaugh, M. A., Boswell, W. R., Roehling, M. V., & Boudreau, J. W. (2000). An empirical
    examination of self-reported work stress among U.S. managers. *Journal of Applied Psychology,
    85*(1), 65–74. doi:10.1037/0021-9010.85.1.65.
Curtaz, K., Hoppe, A., & Nachtwei, J. (2015). Bewusste Auszeiten vom Smartphone tun gut! Eine
    Interventionsstudie zeigt die Wirksamkeit der ("Offtime"-)App in Hinblick auf Erholung und
    Arbeitsengagement. HR Performance (1), 112–114.
Dabbish, L. A., & Kraut, R. E. (2006, November 4). *Email overload at work.* CSCW' 06, Banff,
    Alberta, Canada.
David, K., Bieling, G., Bohnstedt, D., Jandt, S., Ohly, S., Roßnagel, A., et al. (2014). Balancing the
    online life: Mobile usage scenarios and strategies for a new communication paradigm. *IEEE
    Vehicular Technology Magazine, 9*(3), 72–79. doi:10.1109/MVT.2014.2333763.

Derks, D., ten Brummelhuis, L. L., Zecic, D., & Bakker, A. B. (2014a). Switching on and off …: Does smartphone use obstruct the possibility to engage in recovery activities? *European Journal of Work and Organizational Psychology, 23*(1), 80–90. doi:10.1080/13594 32X.2012.711013.

Derks, D., van Mierlo, H., & Schmitz, E. B. (2014b). A diary study on work-related smartphone use, psychological detachment and exhaustion: Examining the role of the perceived segmentation norm. *Journal of Occupational Health Psychology, 19*(1), 74–84. doi:10.1037/a0035076.

Derks, D., van Duin, D., Tims, M., & Bakker, A. B. (2015). Smartphone use and work-home interference: The moderating role of social norms and employee work engagement. *Journal of Occupational and Organizational Psychology, 88*(1), 155–177. doi:10.1111/joop.12083.

Frese, M., & Zapf, D. (1994). Action as the core of work psychology: A German approach. In M. D. Dunnette, L. M. Hough, & H. C. Triandis (Eds.), *Handbook of industrial and organizational psychology* (Vol. 4, pp. 271–340). Palo Alto: Consulting Psychologists Press.

Glavin, P., Schieman, S., & Reid, S. (2011). Boundary-spanning work demands and their consequences for guilt and psychological distress. *Journal of Health and Social Behavior, 52*(1), 43–57. doi:10.1177/0022146510395023.

Hacker, W. (1998). *Allgemeine Arbeitspsychologie: Psychische Regulation von Arbeitstätigkeiten.* Bern: H. Huber.

Hacker, W. (2003). Action regulation theory: A practical tool for the design of modern work processes? *European Journal of Work and Organizational Psychology, 12*(2), 105–130. doi:10.1080/13594320344000075.

Lanaj, K., Johnson, R. E., & Barnes, C. M. (2014). Beginning the workday yet already depleted? Consequences of late-night smartphone use and sleep. *Organizational Behavior and Human Decision Processes, 124*(1), 11–23. doi:10.1016/j.obhdp.2014.01.001.

Lazarus, R. S., & Folkman, S. (1984). *Stress, appraisal, and coping.* New York: Springer.

LePine, J. A., Podsakoff, N. P., & LePine, M. A. (2005). A meta-analytic test of the challenge stressor-hindrance stressor framework: An explanation for inconsistent relationships among stressors and performance. *The Academy of Management Journal, 48*(5), 764–775.

Ohly, S., Sonnentag, S., Niessen, C., & Zapf, D. (2010). Diary studies in organizational research. *Journal of Personnel Psychology, 9*(2), 79–93. doi:10.1027/1866-5888/a000009.

Orlikowski, W. J. (2007). Sociomaterial practices: Exploring technology at work. *Organization Studies, 28*(9), 1435–1448. doi:10.1177/0170840607081138.

Podsakoff, P. M., MacKenzie, S. B., Lee, J.-Y., & Podsakoff, N. P. (2003). Common method biases in behavioral research: A critical review of the literature and recommended remedies. *Journal of Applied Psychology, 88*(5), 879–903. doi:10.1037/0021-9010.88.5.879.

Zapf, D., & Semmer, N. K. (2004). Stress und Gesundheit in Organisationen. In H. Schuler (Ed.), *Enzyklopädie der Psychologie: Organisationspsychologie – Grundlagen und Personalpsychologie* (Wirtschafts- Organisations- und Arbeitspsychologie, Vol. 3, pp. 1007–1112). Göttingen [u.a.]: Hogrefe.

Zapf, D., Dormann, C., & Frese, M. (1996). Longitudinal studies in organizational stress research: A review of the literature with reference to methodological issues. *Journal of Occupational Health Psychology, 1*(2), 145–169. doi:10.1037/1076-8998.1.2.145.